CHILDREN OF THE ICE AGE

CHILDREN

OF THE

ICE AGE

How a Global Catastrophe Allowed
Humans to Evolve

STEVEN M. STANLEY

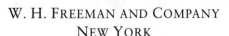

W. H. FREEMAN AND COMPANY
NEW YORK

Cover Image: David L. Brill

Text Design: Cathryn S. Aison

Illustrations © 1996 by Suzanne E. M. Edmonds

Text © 1996, 1998 by Steven M. Stanley. All rights reserved.

Originally published in hardcover by Harmony Books,
a division of Crown Publishers, Inc.
Paperbound edition published by W. H. Freeman and Company.
Printed in the United States of America
First paperbound printing, 1998

Library of Congress Cataloging-in-Publication Data

Stanley, Steven M.
Children of the ice age: how a global catastrophe allowed humans
to evolve/Steven M. Stanley.
p.   cm.
Originally published: New York: Harmony Books, c1996.
ISBN 0-7167-3198-3
1. Human evolution. 2. Glacial epoch. 3. Australopithecines.
I. Title.
[GN281.4.S73   1998]
599.93'8—dc21   97-43543
                                              CIP
                                              AC

CONTENTS

ACKNOWLEDGMENTS

While writing this book, I benefited greatly from the expertise and support of several people. Leslie Meredith and Karin Wood of Harmony Books provided wisdom and advice in the later stages of the book's gestation. Peter Guzzardi did the same earlier, and to him and to my agent, John Brockman, I owe my opportunity to have published with Harmony. Jared Kieling offered exceptionally valuable editorial assistance, especially with aspects of reorganization; thanks to his contribution, my message rings out more clearly. Finally, I owe special thanks to Suzanne Edmonds for providing the excellent artwork and to G. Philip Rightmire for expertly reviewing an early draft of the manuscript.

CHILDREN OF THE ICE AGE

CHAPTER ONE

Spawning a Theory

This is a story about our deep roots. It is about the origin of a small group of closely related species of the genus *Homo*, of which our species, *Homo sapiens*, is the lone survivor.

The story is not simply about names and dates or about the cast of characters in human evolution. It is about why a biological revolution took place. When evolution created early *Homo*, it endowed our ancestors with a powerful brain that elevated them far above all other earthlings in cerebral prowess. The setting for this singular event was Africa, and the time, almost two and a half million years ago. The reasons it came about have only recently come to light.

This is a story that begins with three mysteries. Before *Homo* appeared, its immediate ancestors, members of the much more apelike genus *Australopithecus*, had endured in Africa for more than a million and a half years. During their long stay on earth, they had not evolved significantly in our direction, remaining barely ahead of apes in the size of their brains. That is the first mystery. *Australopithecus* then disappeared, which is the second mystery. It did not vanish without issue, however. During an interval of perhaps a hundred thousand years—but possibly much less— one of its populations evolved into *Homo*. Geologists commonly

measure time in millions of years. By this yardstick events confined
to a hundred thousand years are considered brief, and ten or twenty
thousand years is a mere snap of the fingers. Of course, some catastrophic geological events have been more precipitous. The impact of the asteroid or comet that is now widely blamed for the
extinction of the dinosaurs sixty-five million years ago is thought
to have wrought most of its damage to life on the planet in just a
few days or weeks. Still, the origin of *Homo* was sudden on a geological scale of time, a momentous evolutionary step that is the
third mystery.

This book presents an original explanation for the remarkable
evolutionary stability of *Australopithecus* and the sudden origin of
Homo—two problems that have been widely ignored. It describes
how the onset of the modern Ice Age led to the transition from *Australopithecus* to *Homo* by abruptly altering the African landscape.
Ironically, although a change in the environment caused the extinction of the previously successful *Australopithecus*, it also removed a barrier that had previously prevented this creature from
evolving a much larger brain. Given the new opportunity, one
small surviving group soon emerged from the environmental crisis as an entirely new, big-brained form that had the wherewithal
to make its way in the Ice Age world of Africa: this was early *Homo*,
the product of a difficult birth. In fact, the *catastrophic birth* is my
label for this evolutionary event.

Surprising as it may seem, I will attribute the onset of the Ice
Age—and hence the origin of the human genus—to a regional
geological event that, but for these major sequelae, would seem
trivial when viewed in a global perspective. This event was the uplifting of a narrow bridge of land between two continents far across
the ocean from the African birthplace of *Homo*. As a result of this
modest geological construction, profound changes in oceans and
climates cascaded around the world. The jarring implication of this
chain of causation is that we humans would not exist were it not
for one small vagary of the earth's dynamic crust.

Darwin's Legacy

More than a century ago, Charles Darwin conjectured that Africa had been the crucible of human evolution. His logic could not have been simpler. The Dark Continent is the home of our close relatives, the gorillas and chimpanzees; here, it seemed likely, human and apes shared a common ancestry. Darwin was the first scientist to build a case that an earthly process of evolution had brought forth humans from lower animals, or brutes, as they were known in his day. Although Darwin's notes reveal that he had thought extensively about human evolution long before publishing *On the Origin of Species* in 1859, he skirted the subject in his earthshaking book. In the final chapter he wrote tersely and with profound understatement, "Light will be thrown on the origin of man and his history." Nonetheless, once Darwin had loosed the astounding idea of evolution by natural selection on the starchy, fundamentalist Victorian world, he was immediately set upon by true believers in Adam and Eve who homed in on the notion of a biological link between humans and apes.

Darwin had little choice but to confront human origins directly, but the topic was so touchy that he was slow to take it up in print. In 1871, however, he finally devoted an entire book, *The Descent of Man*, to the subject. Even by that time, there was little fossil evidence bearing on the subject. Scientists had in hand only one partial skeleton of an extinct member of the human family. This was the first find of a Neanderthal skeleton, and prominent anatomists viewed it as nothing more than the remains of an aberrant modern human, perhaps an old man deformed by disease. In theorizing about human ancestry, Darwin necessarily relied heavily on what was known of modern humans and apes.

Embedded in Darwin's view of human evolution were two concepts that pervaded the study of human origins until quite recently. One has become known as the *single-species hypothesis*. This is the idea that only one species of the human family has occupied the planet at any time. In other words, a single line of evolutionary descent has led from apelike ancestors to modern humans. The second, related concept is *evolutionary gradualism.* According to the gradualistic model of evolution, which pervaded Darwin's *On the Origin of Species,* most evolutionary change in the history of life has been wrought by the accumulation of infinitesimal steps of change, generation by generation, over millions of years.

The way in which the single-species hypothesis and gradualism have been applied to human evolution is well illustrated by a cartoon that we can all conjure up from memory. This is the artist's standard rendering of what I call the "up-from-the-apes" view of our origins—the view that modern humans evolved from four-legged walkers by gradually assuming a vertical posture. The typical depiction of this imagined trend shows an ape on all fours at the left and an upright human in full bipedal stride at the right. Between the two figures are alleged intermediate forms that are successively less crouched, such that the last one might almost be someone we have seen on the subway. This hackneyed sequence is not the invention of artists, but of scientists. Even so, it is, quite simply, wrong.

First of all, modern humans did not evolve from animals that habitually moved about on all fours when on the ground. Research over the past twenty years has shown that the structure of the human body and that of *Australopithecus* indicate that our ancestors did not walk on the knuckles of their forelimbs in the manner of modern apes, but instead came down out of the trees on their hind legs. The next chapter explains how this was discovered. The second problem with the cartoon is that it depicts a simple, gradual trend instead of a complex pattern of stepwise changes: it places modern humans at the terminus of an unbranching evolutionary lineage.

Both of these misapprehensions about human ancestry first appeared in *The Descent of Man.* Darwin wrote of human predecessors having become "more and more erect" from the four-legged posture of apes on the ground—having been "gradually converted from a quadruped into a biped." His gradualistic view of evolution prevailed for a century. In 1962, for example, Theodosius Dobzhansky, the greatest experimental geneticist of the mid-twentieth century, published the book *Mankind Evolving*, in which he espoused both gradualism and the single-species hypothesis. He wrote, "The evidence now available . . . is compatible with the assumption that, at least above the australopithecine level, there always existed only a single prehuman and, later, human species. . . ." Dobzhansky went on to claim that slow, persistent transformation, as opposed to rapid branching, "is one of the distinguishing marks of human evolution." Even today, books and articles are published with diagrams that depict a gentle gradient of change between *Australopithecus* and *Homo sapiens*.

A Punctuational Family Tree

The punctuational model of evolution, in contrast to the gradualistic model, proclaims a jerky pattern for evolution, with most change having been concentrated in branching events. According to this model a distinctive new species has typically emerged through the transformation of a small population of a parent species, usually in a marginal environment, during just a few thousands or tens of thousands of years—or even less. The punctuational model was articulated in the early 1970s by Niles Eldredge of the American Museum of Natural History and Stephen Jay Gould of Harvard University. I joined them shortly thereafter, having subjected the model to tests that it passed with strong marks.

Darwin's process of natural selection can easily produce punctuational change. Natural selection, by definition, favors the kinds of individuals that tend to produce many offspring by living long lives or by reproducing rampantly even during normal life spans.

Commonly, a small, isolated population at the margin of a species' range occupies an unusual environment for the species—one that exposes the population to abnormal conditions. Not only may the physical habitat be unusual but so may the food resources, the competitors, and the natural enemies. Such an environment can subject the population to unique pressures of natural selection that cause it to move in its own evolutionary direction—to become a distinctive new species. According to the punctuational model this situation often does more than simply produce a new path of evolution. It speeds up evolution as well by subjecting the entire localized population to intensive selection pressures.

The other side of the punctuational coin depicts stability for well-established species. Once a new species has expanded to become populous and widespread, it has typically undergone relatively little further change, although it may, in its turn, have budded off one or more daughter species before going the way of all species to the oblivion of extinction.

As more and more fossils have turned up, human evolution increasingly has come to look punctuational. The evidence now flies in the face of both the single-species hypothesis and gradualism. As it turns out, our family tree is not a solitary stalk. In fact, it is not even a tree with a central trunk, but more of a spindly bush. It displays branching here and there, where a new species diverged from a parent species and then lived alongside it until one or the other died out. In fact, for much of the time during the past three million years, two or more species of the human family walked the earth at the same time. As the punctuational model would predict, some species survived for long spans of time as relatively stable biological entities, and these periods of relative stability were separated by intervals of rapid evolutionary change. The survival of *Australopithecus* for more than a million and a half years exemplifies stability. The origin of *Homo* near the start of the modern Ice Age exemplifies rapid change.

The pattern of human evolution first attracted my interest

when I was documenting millions of years of evolutionary stagnation for many individual species of animals, plants, and single-
celled creatures. This was in the late 1970s when I was writing a
book that favored the punctuational model of evolution. At the
time gradualism seemed to be more deeply entrenched in physical anthropology than in any other field of evolutionary biology. I
sensed that something was amiss. Even then, I saw the evidence
of fossils tilting the balance toward a punctuational pattern for
human evolution. Today, there is much firmer evidence for the
evolutionary stability of *Australopithecus* and the sudden origin of
Homo. The accumulation of this evidence stands as a conspicuous triumph for the punctuational model: if our own genus suddenly emerged, phoenixlike, from a long-enduring creature of a
very different kind, then even those evolutionists still inclined toward gradualism must sit up and take notice.

Evidence of evolutionary stagnation within another group of
higher primates, the chimpanzees, has recently come from an entirely different scientific venue—genetics. A group led by Phillip
Morin of the University of San Diego has studied groups of chimp
genes that have little adaptive significance but that simply accumulate mutations in the course of time. They found that a population of chimpanzees living in isolation along the west coast of
Africa, between Liberia and Senegal, has diverged quite far, genetically, from the larger group of chimps that occupy the equatorial region to the southeast. The chimps have diverged so far, in
fact, that the western population must have been isolated for about
1.6 million years. Nevertheless, the members of the two groups so
closely resemble each other physically that experts have traditionally regarded the western population as a mere subspecies, or
race, of the common chimpanzee, not a separate species. In spite
of the genetic separation, the two groups have not diverged appreciably in form since they became separated 1.6 million years
ago, and neither group has evolved much at all during this long
stretch of time.

The Terrestrial Imperative

Within a year or so after I had developed a special interest in human evolution, a new idea about human origins jumped out at me—one with important ramifications. It was simply a logical deduction that emerged when I was thinking about one of the many issues which Darwin had contemplated. This issue was the evolutionary timing for the two biological traits that are quintessentially human: our upright posture and our large brain. Which came first, Darwin had asked, our unusual, perpendicular relationship to terra firma or our formidable intelligence?

Lacking fossil data, Darwin approached issues of human evolution, including the timing of brain expansion, by considering what was known about the biology of living apes and humans—and in his day little was known about the apes. (The first scientific description of a gorilla had been published as recently as 1847.) Darwin did, however, note that an ape could not manipulate or carry artifacts much of the time because its forelimbs were often occupied with four-legged walking or climbing. Apes had no need for large brains, he concluded, because they did not normally move about on their hind legs and could not make good use of tools. It seemed to follow that human ancestors must have begun to walk upright before their brains expanded dramatically. Thus, Darwin reasoned, a shift toward upright posture led the way in the evolution of humans. It did so, he thought, by freeing the hands to make and use tools, thus giving selective advantage to individuals who could employ tools most effectively.

Actually, apes do use tools, but only in a very limited way. Modern studies have revealed that chimpanzees are the only modern apes that frequently employ even rudimentary tools in nature. Chimps convey protein-rich ants to their mouths with twigs, squeeze drinking water from leaves that they have wadded into green sponges, and use stones as hammers to crack nuts on larger stones that serve as anvils. They soon discard all of these rudimentary tools, however, except for the large anvils, which they simply put to use wherever they find them naturally positioned on the ground. No modern species of ape manufactures complex artifacts for long-term use.

The idea that struck me in contemplating posture and intelligence was that there is another restriction, more fundamental than the one proposed by Darwin, requiring our ancestors to be fully on the ground, on their hind legs, in order to evolve the large brain of *Homo*. Before the momentous transition could take place, I realized, our ancestors had to have abandoned the simian habit of regularly climbing trees. The essence of this restriction, which I call the *terrestrial imperative*, is that before large brains could evolve the transitional animals needed to have their hands free to carry and tend helpless infants.

Why helpless infants? As it turns out, highly immature offspring are unique to humans, among all primates, and they are a necessary concomitant of the way in which we grow our large brain. The contrast between humans and apes in the rate of maturation is striking. Whereas human offspring remain feeble and uncoordinated for many months after birth, young chimpanzees and orangutans are mature enough to crawl about and cling to their active mothers almost immediately. But it is also during this early interval — in fact, during the first year of life after birth — that humans move far beyond apes in brain size. The result is that a human baby, unlike a young ape, remains top-heavy at the time of its first birthday. A one-year-old child actually has a brain that is more than twice as large as that of an *adult* chimpanzee! As explained in chapter 5, the evolution of delayed development for humans, which makes infants so weak and uncoordinated, is actually responsible for the phenomenal cerebral expansion that takes place during the first year beyond the womb.

I realized that the origin of the human pattern of brain growth amounted to a great evolutionary compromise — one that restricted the activity of the mother. Postnatal helplessness was, in effect, the price that early *Homo* paid for its large brain. From the time *Homo* first evolved, its immature infants have represented excess baggage, quite literally, and this was the key to the discovery I made. Parents have had to tend these fetuslike offspring constantly and carry them from place to place. A pattern of development that entailed such helplessness after birth could never have evolved in a group of animals that habitually climbed trees. Females whose forelimbs were occupied with climbing could not have clutched immature offspring; likewise, the helpless offspring could not have hung onto their mothers. Before human ancestors could evolve the large brain of *Homo*, they had to abandon even the part-time habit of climbing trees. This is the terrestrial imperative.

When I conceived of the terrestrial imperative in 1979, it seemed superfluous because experts already were portraying *Australopithecus* as a fully earthbound biped. This mode of life had been ascribed to Lucy, the famous female *Australopithecus afarensis* whose partial skeletal remains were discovered in 1974. Lucy was supposed to have been as thoroughgoing a ground dweller as we ourselves, and she lived more than a million years before the advent of *Homo*. In other words, experts seemed already to have established the sequence of evolution that I was deducing. First came life on the ground and, much later, the large brain. In fact, my theory seemed to be not only superfluous but also incomplete. I could explain the apparent sequence of events, but not the persistence of *Australopithecus* for such a great stretch of geologic time — nor could I account for the particular timing of the transition. Lacking the major elements of a fuller explanation, I simply sat on my idea, unmotivated to publish.

A few years later, however, new evidence gave my moribund argument new life. As a scientist not specializing in human evolution, I had not been monitoring the current literature in the field carefully, but in the mid-1980s I became aware of powerful new arguments that *Australopithecus* had actually spent a good deal of time in trees. The revised reading of its bones was based both on new fossil discoveries and on reinterpretations of earlier finds. It was now clear that, from its fingers to its toes, *Australopithecus* possessed skeletal features which made it a much better climber than modern humans. As chapter 3 will show, there are good reasons to believe that *Australopithecus* put its climbing skills to frequent use.

The new portrayal of *Australopithecus* changed everything. Suddenly, it was easy to understand why this animal had failed to give rise to big-brained *Homo* for a million and a half years of geological time. If members of the ancestral genus were actually caught between two worlds, one terrestrial and the other arboreal, they could not have evolved into *Homo*. Expansion of their brain in the direction of *Homo* was stifled because *Australopithecus* infants had to cling to mothers that climbed trees as a matter of habit; helpless infants were unsupportable. This crucial point somehow had previously been overlooked.

Of course, *Australopithecus* was ignorant of its prospects. It simply pursued the life that was dictated by its taste for tree-borne fruits and seedpods and by its well-justified fear of ground-dwelling predators. Although it remained stupid by modern human standards, it happily climbed trees, gave birth to infants that could cling to their mothers, and became a highly successful mammal.

The Catastrophic Birth

Knowing why *Australopithecus* failed to evolve a large brain for a remarkably long interval of time leaves open the question why it abruptly evolved a large brain about 2.4 million years ago. Could it be that a pervasive environmental change forced the transition from *Australopithecus* to *Homo*, with its big brain and helpless infants? I believe that this is exactly what happened. A particular idea about environmental forcing came to me some months after I had recognized the terrestrial imperative, and it seemed so obvious that

I was embarrassed not to have hit upon it sooner. The modern Ice Age began not long before *Homo* evolved from *Australopithecus*, and I recognized that this profound geological event could have triggered the transition.

Shortly before *Homo* made its appearance, the natural world differed strikingly from the one that surrounds us today. Mammals of the same general types were present, but the global menagerie was richer, especially in species of giant proportions. Several kinds of large saber-toothed cats feasted on massive pachyderms, not only in the Old World but also in North America, where huge ground-dwelling sloths stood eye to eye with elephants. These sloths were giant cousins of the moss-ridden animals that today hang from trees in South American rain forests. Among the hoofed animals of the New World were a species of camel that was taller at the shoulder than a modern giraffe and a member of the horse family whose fossil bones show it to have been a close relative of modern zebras—so close that we might well imagine that it had stripes.

Then came a profound change in the earth's climate—not global warming of the sort that now confronts us, but widespread cooling and drying. Warm conditions at high northern latitudes were yielding to an ice age, one from which our planet has yet to emerge. This Ice Age has been a time of climatic oscillations. Since its inception glaciers have waxed and waned in the Northern Hemisphere. Today, we are perched precariously in the most recent of a series of temporary glacial retreats. Only the Greenland ice cap and a number of small mountain glaciers now remain, but these frozen white masses flash a warning that in a few tens of thousands of years vast ice sheets will once again plow southward to New Jersey and the Alps. The present Ice Age began with what amounted to a flip of the climatic switch. A hundred thousand years or two before the time of *Homo*, the first large glaciers expanded from northern centers and a chill spread southward across Europe, Asia, and North America. Physical changes that attended the onset of the Ice Age forced widespread biological vicissitudes. As will be spelled out in the following chapters, the changes in Africa, the continent to which *Australopithecus* was confined, were exactly the ones needed to close the door on this apelike creature and to open the way for the genesis of *Homo*.

The idea of a catastrophic origin for *Homo* carries with it a sobering implication: global change is not all bad. We humans, who now fear the ecological consequences of a warming of the planet's climate, owe our very existence to a profound, climatically driven transformation of the earth's ecosystems—one that wiped out many members of the human family but that allowed at least a few to survive by what amounted to an evolutionary rebirth.

How Knowledge Advances

The theory I am offering is not fanciful, but grounded in a variety of factual data. Even so, it is not the immediate, obvious result of a fossil discovery, but rather the product of reasoning that began with a basic deduction, based on the way in which we humans grow our large brain, and then continued over several years through the addition of new facts and ideas. This approach has been unusual for the study of human evolution, where, time and again, it has been the unearthing of a new fossil that has added a key piece to the partially completed puzzle. It has usually been facts, rather than ideas, that have advanced our understanding.

Discovery has jolted conventional wisdom more frequently in the field of human origins than in any other sector of the study of prehistoric life. Serendipity looms large where data are sparse, and fossils that represent early members of the human family are so rare that the facts available at any time have left plenty of room for conjecture about the details of our ancient ancestry. The unearthing of a single fossil has summarily dispatched many an idea that experts previously entertained.

The two central ideas of this book—the terrestrial imperative
and the catastrophic origin for *Homo*—impinge on so many aspects of human ancestry that, to be put forward, they had to square
with many previous fossil discoveries. We may look at the situation in another way: The explanation would have been dead in the
water immediately if it had conflicted with any single piece of a
large body of existing evidence. Its strength is that it logically connects a large variety of conspicuous facts.

Breaking Down Barriers

It was not by design that I entered the field of physical anthropology. I had not set out to explain why the large brain evolved
when it did but simply came to the idea of the terrestrial imperative while reaching out from my primary area of research to contemplate human origins. This exciting idea automatically led me
to consider related issues. Before long, I saw the connection to
global environmental change and formulated the catastrophic scenario. At this point there was no turning back. Finding myself with
one foot in a project that was too exciting to postpone, I plunged
in, setting aside my other research activities. I dug into the literature to extract relevant facts, measured bones, made calculations,
and plotted graphs.

The result of my endeavors was a technical manuscript that I
submitted to the journal *Paleobiology*. One of the anonymous re-
viewers whom the editor of the journal had selected to evaluate
my manuscript dismissed it with the curious statement "There is
little here that is original." When I explained to the editor that both
the terrestrial imperative and the catastrophic scenario, among
other things, were new conceptions, he ruled in my favor. No
doubt he was influenced by the second anonymous reviewer, who
generously wrote, "Who can resist some sense of fascination when
one of our most original thinkers in macroevolutionary theory

turns . . . to questions of human evolution? . . . The paper takes what is known and ties it together in an original, provocative, and reasonable way." *Paleobiology* published my article in the late summer of 1992. There has thus far been little response from anthropologists, although William W. Howells, a retired Harvard professor of anthropology, wrote a kind note saying that he found my paper "very satisfying" and that he regretted having seen it too late to have "drawn on" it for a forthcoming book, the magnum opus of his career. The book was published in 1993.

Holistic Paleontology

The resistance I have faced in venturing into a neighboring scientific field amounts to territorial behavior. Paleontology is my field, and traditionally paleontologists have divvied up the objects they study along taxonomic lines. Specialization inevitably has led to the pigeonholing of animals into much smaller groups for particularized study. In the course of this partitioning, students of prehistoric human evolution—paleoanthropologists—have come to stand farthest from all other specialists, even those who focus on other groups of animals with backbones. Hundreds of scientists have dedicated their careers largely to the study of the fossils of *Homo* and *Australopithecus*. This intense concentration of effort reflects our species' understandable fascination with its own recent evolutionary history. The rapidly evolving culture that sets humans apart from all other animals has promoted the academic separation of those who study *Homo* and *Australopithecus* from other biological scientists. Thus, anthropology has become in part a social science.

It is unfortunate that arbitrary academic boundaries have attenuated the dialogue between students of human evolution and other paleontologists. Now, however, old barriers are falling throughout science. Many of the most exciting recent advances have come at junctures between traditional disciplines. In effect, research has been straddling boundaries, and as a result, the boundaries have become blurred. Established scientists have been retrofitting themselves with new skills, borrowing methodologies and instruments from neighboring fields, and exchanging ideas with other kinds of researchers to achieve scientific cross-fertilization. Paleontology is inherently a hybrid science anyway, occupying the area where geology and biology meet. I sometimes ask myself whether I am a geologist or a biologist. If I can straddle these fields, why should I not bring a fresh approach to the overlapping terrain of anthropology?

Holistic paleontology is my label for the multifaceted study of ancient life, the implication being that the whole is greater than the sum of its parts. Making connections is the key to success in holistic paleontology—connecting the appropriate technique or concept with a problem or identifying a cause-and-effect relationship between two events of the distant past. When I first recognized the connection between the big brain of *Homo* and a fully terrestrial life, my knowledge of anthropology was thin. On the other hand, I benefited greatly from my personal experience in other areas of research. Let me illustrate this point by reviewing some of the basic principles that guided me as I delved into the origin of *Homo* and that I have also put to use in the study of lower forms of life.

Thinking of Humans as Animals

I first called attention to the evolutionary stagnation of *Australopithecus* and the rapid origin of *Homo* in my book *Macroevolution: Pattern and Process*, published in 1979, long before I had conceived of the terrestrial imperative or the catastrophic scenario. As I saw it, the pattern of human evolution squared with the punctuational model, which my book supported generally for animals, plants, and single-celled creatures. My numbers actually indicated that *Australopithecus* was not an exceptional form of mammal in surviving for more than a million years with little change, but rather a typical one. At the time of my writing, a gradualistic view of human ancestry prevailed. Had I been a confirmed gradualist at the time, rather than something of a rebel, I would not have been predisposed to identify a reason either for the stability of *Australopithecus* or for the abrupt emergence of *Homo*.

The picture that I have sketched of the origin of *Homo* credits large African predators with having played a prominent role in the early evolution of the human family. They compelled our ancestors to spend part of their lives in trees until the Ice Age began. Most anthropologists have had a different focus. They have concerned themselves more with what our ancestors ate than with what ate *them*. Evidence of this emphasis is the long-enduring popularity of the single-species hypothesis—a model founded on the belief that species of the human family should have tended to compete so intensively for food or territory that the world could not support two of them at a time. If evolution ever produced a new branch of the human family, the single-species hypothesis asserted, extinction should quickly have pruned either this branch or the original stem. When the two species came into contact, one should soon have eliminated the other through competition for food or territory. In other words, the African continent, our ancestral home, was not supposed to be big enough for two of us.

When first proposed, the single-species hypothesis w~~ ~~ grounded in contemporary ecological '¹ stand that two very similar species can ε out one defeating the other. All that is ¹ disease, or environmental disturbance k₁ perior competitor in check. A species tha₁ cannot monopolize resources such as ᵗ

leaves room for other similar species that are less effective competitors to survive in the same area, utilizing whatever resources
remain unexploited.

My first serious observations of life in the wild were in shallow
seas, and it was there I learned about the power of predation. I wal-
lowed in mudflats, dredged animals from boats, and dived on
reefs. I soon came to recognize that heavy predation alleviated
competition by suppressing the populations of potentially dominant species, ones that would otherwise tend to dominate others.
My subjects were bivalve mollusks—clams, scallops, mussels, and
their relatives—and I discovered that predators make a habit of dec-
imating populations of these animals. Pick up ten or fifteen small
clamshells from a sand flat at low tide and you are likely to find
that a large percentage are perforated by a neat round hole, usually positioned near the middle, where the main body of the animal was housed. Such a hole is the mark of a carnivorous snail that
bored through the shell with a mechanical drill assisted by secre-
tions of acid. The snail's tubular proboscis entered the hole, turning the clam into a succulent meal. Shell-cracking crabs also take
a heavy toll on clams. Bivalves are so heavily preyed upon that their
populations are typically sparse and unstable. The result is that they
do not vie intensively for food or space, and many similar species
can live together, sharing the same resources. So, too, should two
or more species of the human family have been able to share the
African continent in times past.

Reading the Shapes of Shells and Bones

Early in my career, my research focused on form and function.
studied the ways in which evolution molded skeletal shapes of
lve mollusks to suit particular ways of life. I scrutinized the be-
al ecology of nearly a hundred bivalve species in the field
atory. I spied on buried clams with X rays, made robots

that imitated the burrowing movements of live animals, and tested the effects of waves on animals half-buried in sand within a ten-foot aquarium that rocked on an axle. Through observations and experiments, I unraveled the meaning of scores of shell features. It turned out that certain types of spines and ridges help clams to burrow into sand, whereas other types anchor them there, unequal shell halves turn scallops into hydrofoils when they swim by jet propulsion, and huge teeth along the hinge of a cockle prevent misalignment of the shell halves when the animal jumps with its muscular foot to avoid a starfish or crab that threatens it. This research allowed me to interpret the fossil record of an entire class of animals and to reconstruct its general evolutionary history.

The internal skeleton of a human bears little resemblance to the external skeleton of a mollusk, but the two are studied in compliance with the same principles of mechanics and evolution. We learn from evolutionary theory, for example, that natural selection must operate without the freedom of an engineer. For one thing it must work its effects on existing forms of life. It cannot create a species de novo from raw materials or even from small parts. Evolution is in the business of remodeling, not new construction, and even some forms of remodeling are more probable than others.

Early in my studies of clams and their relatives, it became apparent that they were far from perfect, owing to these limitations of evolution. To appreciate the sources of imperfection, imagine that a clever biological engineer of the future designed the most beneficial evolutionary changes that he could imagine for several species of the modern world and then boarded a time machine to see whether, during the next few million years, evolution had followed his plans. He would be sorely disappointed. Evolution would probably have produced new, well-adapted forms of life, but none that closely resembled those he had designed. This situation is reminiscent of what is amusing about the reply of the person asked to give directions: "You can't get there from here!" Seldom can one say that anything is truly impossible. There is always some way to get from point A to point B. Sometimes, however, every conceivable path is so difficult that the trip is *almost* impossible. Likewise, it is often next to impossible for evolution to transform one kind of organism into another form that might look good on the drawing board. Because an existing species embodies complexly interwoven structures and physiological processes, any truly drastic change in its intricate pattern of growth is likely to be disruptive—in fact, fatal. Useful changes that are less wrenching will be more likely to evolve, but even these are apt to have negative side effects that handicap their biological productions. We humans are certainly the flawed productions of an imperfect natural process. It would be difficult to find a better example of evolutionary compromise than the trade-off between the evolutionary origin of the huge brain of *Homo* and of the infantile helplessness that accompanied it.

Timing Is Everything

Much of evolution amounts to a change in the pattern of development, or maturation, of an organism. Often the change is in the timing of developmental events. The origin of the big brain of modern humans illustrates this principle. Chapter 5 will show how human ancestors evolved their large brain by undergoing a delay in their rate of maturation; they grew in height and weight just as rapidly as before, but they simply matured more slowly.

I have studied a similar phenomenon in the lowly bivalves, and for them, too, a change of timing had enormous ecological consequences. A particular kind of evolutionary delay endowed adult mussels and their relatives with a byssus, the bundle of threads by which they attach to rocks. The byssus originally evolved more than four hundred million years ago as an organ of attachment for juvenile clams that lived on the seafloor. It anchored them to grains of sand, when they would otherwise have been at the mercy of waves and currents. They soon outgrew the byssus, however, and lived free. Nearly half a billion years ago, certain types of burrowing clams evolved in a new direction. The shapes of their fossilized shells reveal that they retained the byssus into adulthood, remaining attached to grains of sediment throughout their lives. This evolutionary innovation opened the way for some bivalves to emerge from the sediment through further evolution and to cling to rocks and reefs throughout their lives. Thus were born the ancestors of modern mussels and scallops, and life of the seafloor was changed forever, much to our culinary benefit.

In just such a way did a delay in our ancestors' development lead them to a new way of life.

The Ice Age on the Land and in the Sea

The ecological crisis that ushered *Homo* into the world also has its counterpart in the marine venue of my previous work. I would probably not have recognized what happened to human ancestors in Africa had I not spent so many hours with pick and trowel in the Atlantic Coastal Plain, which is a treasure trove of fossil seashells. Many of the specimens two or three million years old remain intact and shiny, looking as if they might have housed living animals just a year or two ago before being washed up on a modern beach. A few years ago, my research on these ancient shells revealed that the onset of the recent Ice Age had caused a mass extinction of marine creatures throughout the North Atlantic Ocean. The agent of this destruction was the cooling of climates that occurred while glaciers grew on lands that bordered the North Atlantic. The refrigeration spread even to the tropics, so that species of bivalves and snails disappeared by the hundreds from the shallow seas that fringed what is now southern Florida. My discovery of this crisis engaged my interest in other events of the Ice Age and put me in touch with people such as Raymonde Bonnefille, a

French student of plant fossils who had previously shown how certain African environments were transformed at the start of the Ice Age, and Elisabeth Vrba of Yale University, who had shown that such changes somehow triggered a transformation of the antelope fauna throughout Africa. At this time I was unaware that in 1975 Vrba had suggested the possibility of a link between these events and human evolution. Nonetheless, I was well positioned to envision a particular kind of connection myself.

Solving the Mysteries

The link between the Ice Age and human evolution is, in fact, the solution to the second and third of the mysteries that were posed at the start of this chapter: the extinction of *Australopithecus* and, before its demise, the abrupt origin of *Homo* from one of its populations. To tackle the second mystery, in chapters 4 and 5 I will explore the Ice Age and its impact on the world of *Australopithecus*. In chapters 6 and 7 I will take on the third mystery by examining the human brain and asking how the onset of the Ice Age should have allowed this remarkable organ to evolve. In chapter 6 I will trace the histories of the various species of *Homo* that then arose and that invented increasingly complex cultures with increasingly advanced brains.

First, however, it is necessary to step back and investigate the everyday life and death of *Homo*'s immediate ancestor, *Australopithecus*, in an African landscape that was prowled by packs of savage predators. Only then will we be equipped to unravel the first mystery—the persistence of *Australopithecus* with little change for more than a million and a half years. Solving the first mystery will set the stage for addressing the second two, which add up to the question, Why did the apelike *Australopithecus*, after such a long stay on earth, give way so abruptly to *Homo?*

CHAPTER TWO

The Southern Ape

In 1924 a student presented Raymond Dart of the University of Witwatersrand in Johannesburg with the fossil skull of a baboon collected near the South African town of Taung. The fossil was from the rubbly rock within an ancient cave that had been exposed in a quarry, and Dart, sensing an opportunity, ordered up a large volume of the parent material. As he rummaged through the resulting shipment, something remarkable caught his eye. It was the lumpy sedimentary filling, or natural mold, of the interior of a skull. Dart searched for the block from which the mold had broken; this block might sequester the skull itself. He eventually found the block, recognizing a hollow in its surface that matched the mold, and after weeks of chiseling and picking, he had expanded a few bits of exposed bone into one of the most important fossil finds in the history of anthropology. What Dart uncovered was soon to be known as the Taung skull.

One thing about the Taung skull was immediately clear: it represented a juvenile animal that had died while its front molars were in the process of emerging. Otherwise, the skull was a cipher. The braincase was not much larger than that of an ape, but the configuration of the base of the skull showed that, in life, the head had perched atop an upright spinal column. In other words, the skull represented an ape-brained animal that stood upright, like a human. Its antiquity was unknown.

In a paper published in 1925, Dart described his find and assigned it to the new genus *Australopithecus*, "the southern ape." Despite choosing this name, he portrayed the strange creature as having been intermediate between apes and humans—"ultra-simian and prehuman."

Most anthropologists of Dart's day rejected his proclamation that the Taung skull was a "missing link." Instead, they viewed it as some sort of juvenile ape that had no special ancestral relationship to humans. One obstacle to viewing it as a human ancestor was the existence of a fossil skull that had been discovered in 1891. This skull belonged to a species nicknamed "Java man," now formally labeled *Homo erectus*. Today we know that the form ranged over a broad geographic area, but the site of its first discovery engendered a widespread belief that humans had evolved in the region of Indonesia and Malaysia. This idea ran counter to Charles Darwin's opinion that humans had originated in Africa, the land of gorillas and chimpanzees. Darwin had been right, of course, but his opinion remained in doubt at the time of the Taung skull's discovery.

A second problem was that some fossil monkey skulls collected from the Taung site were quite similar to those of living baboons. This circumstance suggested that the whole fossil assemblage was quite young—too young, it was argued, for the Taung creature to be an ancient intermediate between apes and humans. In a remarkable display of elitism, Arthur Keith, curator of anthropology at the British Museum, suggested in a letter to *Nature* magazine that "a genealogist would make an identical mistake were he to claim a modern Sussex peasant as the ancestor of William the Conqueror." We now recognize that the argument based on timing was unfounded. It turns out that the Taung skull is actually at least 2.4 million years old and that some fossil skeletons representing African mammals of the same vintage closely re-

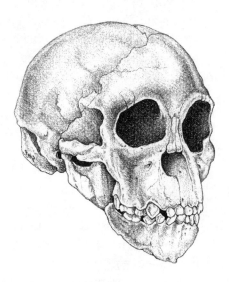

FIGURE 2.1

The Taung skull. This remarkable specimen, the first fossil of *Australopithecus* ever discovered, belonged to a three- or four-year-old juvenile. The rear portion of the braincase, which is missing, is represented by a filling of sediment that records the configuration of the inner surface of the skull.

semble those of living species. Evolution has generated some remarkable new species since that time, but it has left many forms of life relatively unchanged. In other words, between then and now, the baboons whose skulls were found with the Taung skull may not have evolved very much, and yet the species to which the Taung skull belonged could have budded off distinctive new members of the human family. All of this would be quite compatible with the punctuational model of evolution.

Other mistaken ideas about evolutionary timing also stood in the way of Dart's proposal that *Australopithecus* was a missing link between apes and humans. Anthropologists of the day had largely abandoned Darwin's arguments about the general sequence of events in human evolution. In the 1920s few believed that upright posture—the posture of the Taung child—had preceded a big brain. Arthur Smith Woodward, keeper of paleontology at the British Museum, argued that human dominance of the modern world implied that cerebral expansion led the way in our ascent from apes. This claim was specious, quite simply because our present status says nothing about the path by which we got where we are.

Woodward's position reflected a human-centered attitude that was widespread among anthropologists of his day. If they no longer saw the human mind as the crowning achievement of divine creation, they nonetheless viewed it as the pinnacle of organic evolution. Some actually believed that, once natural selection had produced an ape, a mysterious, inner driving force had taken over, expanding the brain inexorably toward the human condition.

The infamous Piltdown hoax also inveigled some anthropologists into believing that the brain had led the way in human evolution. The Piltdown skull, which had purportedly been discovered in a gravel pit in 1908, seemed to indicate that the large human brain had emerged in an otherwise apelike cranium, one with a projecting face and big front teeth. It was nothing but a clever forgery, however, whose perpetrator remains uncertain even today. The skull was actually that of a modern human, which some scoundrel had fitted not simply with false teeth but with an entire false jaw—that of a present-day orangutan. No part of the structure was a real fossil, but until its inauthenticity came to light, some anthropologists gave it credence as a human ancestor. Influenced by this "discovery," they viewed the Taung skull as a remnant of an ape, because not only its jaw but also its brain was much more apelike than human. The Piltdown skull seemed to prove that expansion of the brain, by itself, had led the way toward the human condition.

Other experts had a different theoretical argument for denying that the Taung animal was intermediate between apes and humans. They insisted that, in molding humans from apes, evolution should have transformed all bodily features simultaneously. To them it seemed inconceivable that either the head or the body would have lagged behind in the evolution of the human condition. Believing that all anatomical traits should have become humanized in concert, they attributed the Taung skull to a species off the main line of human evolution: it was nothing but an ape with a strange, upright posture.

In time, however, Dart was vindicated. Since his discovery of the Taung skull, a succession of fossil finds has given scientists a remarkably complete picture of the general anatomy of *Australopithecus*. This is the picture of a creature that was in many ways intermediate in form between apes and humans. It was, indeed, our ancestor.

The Temporary Rise of Gradualism

Unfortunately, the restoration of *Australopithecus* to the status that Dart had favored—the status of missing link—led scientists in yet another wrong direction. They assigned it a gradualistic role in human ancestry. In one published genealogy after another, *Australopithecus* appeared as a segment of a gradually evolving lineage that culminated in *Homo sapiens*.

I will describe quite a different evolutionary position for *Australopithecus*. As the picture of this creature has come into clearer focus during the past several years, crucial new features have come to light. Ironically, our image of *Australopithecus* has changed far more than did the creature itself during its long life in Africa. As noted in chapter 1, *Australopithecus* evolved hardly at all in the direction of modern humans for a hundred thousand generations or more.

Who was this long-lived human ancestor? What way of life did it pursue for so long before yielding abruptly to big-brained *Homo*? These will be the primary subjects of the rest of this chapter. First, however, a brief digression is in order to consider how it is that scientists are able to position fossils in geologic time in order to piece together the story.

Dating Old Bones

For years experts have recognized a southern and a northern species of *Australopithecus*. *Australopithecus africanus*, which the Taung skull represents, is still known only from South Africa. Bones of the northern form, *Australopithecus afarensis*, have come from Kenya, Tanzania, and Ethiopia. As far as is known, the northern form is the older. Its collected fossil remains range in age from 3.9 to about 3 million years in age, whereas those of the southern form span the interval from about 3 to 2.6, or possibly even 2.4, million years ago.

Ages of fossils, measured in millions of years, come from the study of naturally occurring radioactive isotopes—atoms that contain a number of neutrons that makes them unstable, so that they decay by the emission of subatomic particles. European scientists discovered radioactive decay shortly before 1900 and quickly recognized that it fueled the earth's inner furnace, providing the energy that thrusts mountains skyward. Geologists soon came to understand that naturally occurring radioactive minerals would allow them to date rocks. In 1913, at the young age of twenty-five, the great British geologist Arthur Holmes reviewed the incipient technique in the small book *The Age of the Earth*, where he concluded, "Radioactive minerals, for geologists, are clocks wound up at the time of their origin." Today, what we call radiometric dating of rocks provides dates that extend back nearly four billion years.

A rock's age is calculated from the known rate of decay of a particular isotope—the rate of emission of the kinds of particles that make a Geiger counter click. In the simplest case, in order to calculate the age one simply needs to measure the concentration of the radioactive isotope in the material to be dated and also the concentration of the daughter product—the isotope produced by the decay. Radiometric dates for volcanic rocks provide most ages for fossils of *Australopithecus* and early *Homo*. The analytical results are always slightly in error, but larger uncertainties about the ages of fossils result from the simple fact that fossils are not generally found within the volcanic rocks that are dated but in strata above or below them.

Using carbon 14, the most famous radioactive isotope, it is possible to date a wide variety of organic objects, including bits of wood, charcoal, and cloth—even individual seeds! The limitation of this so-called radiocarbon method of dating is that carbon 14 decays so rapidly that it provides ages only for objects younger than about fifty or sixty thousand years. After sixty thousand years or so, only about .1 percent of the original amount of carbon 14 remains—too little for precise measurement. Of the more slowly decaying natural isotopes, potassium 40 is especially useful for dating early members of the human family. Its daughter product is argon 40, which, though a gas, remains trapped within the naturally occurring crystals where it forms.

In recent years new methods of measuring degrees of radioactive decay have greatly expanded our ability to date strata. Furthermore, geologists can assign ages even to fossils collected many miles from rocks that have been dated radiometrically. Rock magnetism offers one method of applying a single date to strata in far-flung regions of the globe. When lava that is rich in iron cools to form solid rock, many of the newly forming crystals become magnetized in alignment with the earth's magnetic field. In effect, such a rock is a natural compass whose magnetism is easy to measure. Sedimentary rocks that form when particles settle from water are similarly magnetized if some of their particles contain iron. Even if only slightly magnetic, these grains tend to align with the earth's magnetic field when they accumulate. What turns the so-called remnant magnetism of the resulting rocks into a tool for dating rocks is the occasional reversal of the earth's magnetic field. This field, which causes a compass needle to point north today, is generated by the motion of the hot, iron-rich liquid that forms the outer portion of the earth's core. At irregular intervals this motion changes and the earth's magnetic field reverses itself. Overall, about half of the time the field has had its present, "normal" polarity. The rest of the time, when the polarity has been reversed, a compass needle would have pointed south.

Over the years geologists have constructed a detailed chronology of magnetic reversals for the entire Age of Mammals. A local sequence of magnetic rocks that formed during a portion of this era can display a characteristic pattern of normal and reversed intervals, some short and some long. When such intervals are plotted as alternating black and white bands on a column that represents geological time, the result resembles a bar code on a piece of merchandise and can be read in much the same way. Such a sequence of magnetic rocks forms the walls of Olduvai Gorge in Tanzania, the site where Mary and Louis Leakey made many of their famous early fossil finds. As a result of these and subsequent discoveries, Olduvai has proven to be the richest trove of anthropological treasures of the past two million years.

Olduvai Gorge is a deep chasm in the Serengeti Plain, a savanna famous for its magnificent mammals. The stream that has incised the gorge flows eastward into one of the great rift valleys of Africa. The rift valleys are sinuous basins bounded by faults, between which the earth's crust has subsided. They trend southward more than two thousand miles from the Red Sea and Gulf of Aden, which themselves occupy rifts that cleaved Saudi Arabia from Africa just a few million years ago. For more than two thousand miles, the rift valleys trace out a zone along which cataclysmic forces are straining to tear Africa apart. Whether the incipient rifts will actually rend the continent and form a new ocean remains to be seen. The continent has resisted the diverging forces for some twenty million years, and its fate will not be settled until many more millions of years have passed. The unanswered question is whether a shift in the pattern of movements deep within the earth will end the rifting process before East Africa separates from the rest of the continent to become a large island in the Indian Ocean.

Earth movements within Africa exemplify a process that affects the entire globe. The outer shell of the planet is divided into large plates that move in relation to one another, but the pattern of movements changes from time to time. The African plate includes not only the continent whose name it bears but also neighboring segments of the Atlantic and Indian Ocean basins. More than two hundred million years ago, a gigantic continent broke along a jagged rift and North America parted from Africa and Eurasia. Thus was formed the Atlantic Ocean, which continues to grow as molten rock, known as magma, wells up from the mantle of the earth and hardens into new seafloor along the Mid-Atlantic Ridge, pushing apart the Old and New Worlds, along with the seafloor in between. At about the time the Atlantic Ocean began to form, rifting to the east began to create the Mediterranean basin, separating Africa from Eurasia. Later, in the region of the Middle East, Africa pressed against Europe and Asia once again. By five million years ago, however, the land bridge had become restricted, and large primates were unable to negotiate it for several million years thereafter. During this interval of geographic isolation, Africa became the crucible of early human evolution.

Scientists owe a great deal to the subsequent rifting event that is still threatening to fragment Africa. It has provided us with a record of human evolution in the lowlands of the rift valleys, where sand and mud have accumulated in and around lakes and rivers, entombing the bones and artifacts of our ancestors. The strata that form the walls of Olduvai Gorge display an exceptionally complete record of the earth's flip-flopping magnetic field, and some of these strata harbor fossils. In fact, one reversal, which switched the magnetic pole from south to north about 1.8 million years ago, was documented there for the first time and is named the Olduvai event. The next reversal, also frozen into the rocks at Olduvai, took place about 200,000 years later. Olduvai has served as one of the best localities for giving the bar code of the earth's magnetism a proper time scale for the past 3 million years because sandwiched between the strata, at intervals, are volcanic rocks that lend themselves to potassium-argon dating. Most of the datable rocks are hardened volcanic ash, inelegantly termed tuff. This ash spewed from volcanoes that grew along the rifts where molten rock ascended from the earth's mantle. One of these East African volcanoes is the famous Mount Kilimanjaro, now a quiescent, snowcapped peak. The volcanism has not ended, however, but will continue as long as the rifts remain active.

Even without radiometric dating, fossils allow scientists to establish the relative ages of rocks. When abundant fossils represent several species in a particular region, we can often see that some of the species existed before or after others. Observations of this type allow paleontologists to array the durations of species in relation to one another. Sporadic radiometric dating then supplies the fossil occurrences with actual ages, sometimes with substantial uncertainty but often with great accuracy. The Taung skull, once assumed to be quite young, is now dated to approximately 2.4 to 2.6 million years. This age comes not from radiometric dating of the Taung rocks themselves, which contain no suitable isotopes, but from fossils of other mammals found with the Taung skull. The approximate intervals during which the other species lived are known from fossil occurrences elsewhere in strata of known age.

By the application of such methods, we know that the collected fossils of *Australopithecus afarensis*, the northern species, range in age from about 3.9 to 3 million years, whereas the known remains of *Australopithecus africanus* date from about 3 to 2.6 or

2.4 million. Thus, the Taung "child" was young in more than one sense—in its age at death and also in its geological antiquity. In 1994, Tim White of the University of California at Berkeley and his coworkers reported on fragmentary remains of the human family in northern Africa that are about 4.4 million years old. These teeth and bones represent a primitive species that is in some ways intermediate between *Australopithecus africanus* or *afarensis* and an ape. They have placed this very ancient species in the new genus *Ardipithecus*.

The year after *Ardipithecus* was described, a group of workers led by Meave Leakey reported on a new, very early species of *Australopithecus* from the Lake Turkana region of Kenya. They named it *Australopithecus anamensis* and assigned its various teeth and bone fragments ages ranging from 4.2 to 3.9 million years. The fossils yield limited and somewhat puzzling information about the biology of the new species. Its shallow palate and large canine teeth have a primitive look. On the other hand, a partially preserved shinbone is quite large and resembles that of *Australopithecus afarensis*; it clearly represents an animal that spent time on the ground. The latter species, to which Lucy belongs, may have diverged from *Australopithecus anamensis* close to 3.9 million years ago, when the two species appear to have coexisted. In any event, from soon after the time when *Australopithecus afarensis* appeared, the genus *Australopithecus* persisted for about 1.5 million years without substantial evolutionary change. Given this view of the australopithecines' chronology, let us take a closer look at its biology.

What Exactly Was Australopithecus?

So similar are *Australopithecus africanus* and *Australopithecus afarensis* that even in the 1980s some experts assigned them to a single species. The resemblance justifies description of the general features of *Australopithecus* without concern for the minor differences between the northern and southern forms. Much research has been undertaken to reconstruct the biology of these forms—to "bring them back to life." Early members of the genus *Homo* were the "after" in a profound evolutionary transformation, and *Australopithecus* was the "before." Let us first have a look at *Australopithecus* above the neck.

Many parts of the skeleton of *Australopithecus* below the neck are more human than apelike, but this is not true for the skull. A skull's cranial capacity is a good measure of the weight of the brain that was housed within it. On this score *Australopithecus* lives up to the literal translation of its name—the southern ape. Modern humans are far brainier. Males of the human family, being larger than females, require larger brains to operate their organ systems. For each gender of modern humans, some 95 percent of all adults have brains within about 20 percent of the average size. This puts 95 percent of females between about 1,100 and 1,340 cubic centimeters in brain volume and the same proportion of males between about 1,280 and 1,560 cubic centimeters. In other words, each gender is quite variable. As it happens, there is only a weak relationship within the human species between brain size and standard measures of intelligence: many very smart people have relatively small brains. Even so, a large difference in *average* brain size between any two species of primates, when adjusted for body size, almost certainly indicates a difference in intelligence.

Scientists have in hand only about five adult skulls of *Australopithecus africanus* that are well enough preserved to have yielded good estimates of cranial capacity, and even fewer skulls of *Australopithecus afarensis*. In addition, there is no way to tell for certain whether any isolated skull belonged to a male or a female. The estimated volumes for the southern skulls range from about 438 to 485 cubic centimeters. The males presumably averaged slightly more than 450 cubic centimeters, and the females, slightly less.

Chimpanzees and orangutans closely resemble one another in adult body and brain weights. Their average cranial capacities are about 430 cubic centimeters for males and 350 for females, but measured values for males range up to about 500. This means that some members of Australopithecus were within the upper range for male chimps and orangs, although these members of the extinct species may have been small females. In any event, Australopithecus, which was slightly below chimps and orangs in average body weight, was only slightly above them in brain size. In both body and brain weight, it was much closer to apes than to modern humans. We are only moderately heavier in total body weight than Australopithecus, yet in brain size we outweigh it by a factor of three. In fact, Australopithecus had a brain about the size of an orange.

For teeth the story is somewhat different. Humans have evolved much smaller front teeth than those of apes, but our molars have become larger. The front teeth of Australopithecus remained large—nearly as large as those of a chimp. On the other hand, its molars had expanded dramatically; they were even larger, relative to body size, than ours and were more thickly enameled. The large front teeth probably served for biting into fruits, as they do in apes. Natural selection presumably expanded the molars and thickened their enamel because nuts and fruit pits, which require crushing, formed a large part of the australopithecine diet.

The shape of its teeth leaves no doubt that Australopithecus was primarily a vegetarian. Like modern chimpanzees, however, it probably supplemented its diet of fruits and nuts by consuming small mammals. Fruit makes up about two-thirds of the diet of chimps, and leaves and seeds, another 20 percent. The British ethologist Jane Goodall was shocked to discover that chimpanzees not only add insects to their largely vegetarian diet but also the meat of the mammals they kill. About half of their victims are primates. These are mostly monkeys, including baboons, but chimps occasionally practice cannibalism, and they have even been known to devour human infants. Their mammalian diet also includes rats and mice, young bushpigs, and bushbuck fawns. Chimps value meat highly, and they sometimes contest each other for a share of a kill. Perhaps meat is especially precious to them because they are not very talented hunters. To improve their efficiency, they often hunt in groups. It is not uncommon for those that make a kill to share meat with others, and the recipients often include supplicants that beg for a handout. Whether *Australopithecus* engaged in similar social behavior we will never know, but the odds are good that it occasionally consumed small mammals.

Large, Posturing Males

Surprising as it may seem, important aspects of social behavior can be read from the reconstructed weights of extinct primates. *Australopithecus*, on average, would have tilted the scales at slightly less than a modern chimpanzee, but this simple comparison masks an important pattern. In the extinct genus males were much larger than females. Henry McHenry of the University of California at Davis recently estimated weights for members of the northern species from the sizes of their hind limb joints. Females, he concluded, averaged about 30 kilograms (66 pounds) and males about 45 kilograms. In other words, males were half again as large as females. In contrast, modern men are only about 25 percent heavier than modern women.

The great disparity in body size between males and females suggests something important about the sex life of *Australopithecus*. Males were probably polygamous, with a few dominant animals monopolizing the females within a troop. Polygamy is the rule for nearly all living primate species in which males are at least 50 percent heavier than females. The explanation lies in what Darwin termed *sexual selection*.

When Darwin contemplated the reality of natural selection,

his attention was drawn to the great differences that males and females of many species display in size, shape, and color. In general,
if natural selection had produced the great diversity of life, it must
somehow have produced what we now call sexual dimorphism, or
differences between the two genders in bodily features. Some such
differences reflect contrasting modes of life, but for others Darwin
recognized that sexual selection had been at work. This is the
accentuation of features that endow individuals with greater than
average reproductive success for their gender. Thus, Darwin explained the development of the male peacock's plumage, which
attracts not only human admirers but also peahens. The more
elaborate the plumage, the greater the males' appeal to females
and, as a rule, the more offspring it will sire.

Large body size for males is the most common product of sexual selection in the animal world. Large males tend to defeat
smaller ones in competition for females when physical combat or
intimidation is involved. Other factors, such as a need to climb
trees or survive when food is scarce, obviously limit the sizes of
males that vie for mates. Even so, the results of sexual selection are
sometimes spectacular, as for elephant seals, in which males average more than three tons, four times the weight of females. It is
possible that males of *Australopithecus afarensis* evolved to be
about 50 percent heavier than females because they differed from
their mates in locomotory behavior, but what we know of modern
primates makes sexual selection a much more likely source of the
discrepancy. Males probably vied for females, and harems or
promiscuous liaisons, not monogamous common-law marriages,
were the norm.

Early Evidence of Two-Legged Progression

If the base of the Taung skull suggested that its owner stood
upright, a large portion of a fossil pelvis that Raymond Dart de-

scribed in 1949 left no doubt. In several ways it was shaped like ours. An ape's pelvis is elongate, extending far up the back of the animal. The human pelvis, in contrast, is a basketlike structure that functions for balance as we walk or run. It basically helps to solve the problem of supporting our body as we shift our weight from one leg to the other. When we walk or run, our center of gravity is not above either foot but in between. This means that when we take a step with the right foot, for example, the muscles of the right hip must keep us from toppling to the left. These muscles attach to the broad right flange of our pelvis. The pelvis of *Australopithecus* resembled ours, although, for unknown reasons, it was narrower from front to back. This was the pelvis of an animal that moved about on two legs.

The physique of the northern species, *Australopithecus afarensis*, is well known thanks to Donald Johanson's discovery in 1974 of the skeleton he christened Lucy. Lucy's bones are also clearly those of a bipedal animal. Her collected remains, which constitute more than 40 percent of the original skeleton, include not only a large portion of her pelvis but also numerous limb bones. Those of the upper and lower leg met at the knee in much the way that ours do. Whereas the hind limbs of an ape are normally almost straight when extended, Lucy was slightly knock-kneed, like a typical modern human. Our thigh bones angle slightly inward as they extend downward from hip to knee. This puts our knees and also our feet close together. The result is that our feet are more nearly beneath our center of gravity than they would be if our legs simply extended straight down from the hip to the ground. The angle at which Lucy's limb bones met at the knee shows that she shared with us this adaptation to two-legged locomotion. Her foot was also arched like ours, acting as a spring when she took a step. Apes, in contrast, have flat feet. All of these features point to a two-legged pattern of progression for *Australopithecus*.

A remarkable fossil find at Laetoli, in northern Tanzania, provides direct evidence that *Australopithecus* spent at least some of its time walking upright. Working there with Mary Leakey in 1978, Paul Abell came upon sets of ancient footprints that almost certainly represent the human family—entire trackways turned to stone. The prints were left in newly settled volcanic ash, which hardened after being wetted by rain that left telltale pockmarks on the surface. These footprints are not alone. Many other kinds of tracks also mark the hardened ash, some attributed to a waddling guinea fowl and others to a squatting spring hare, a large saber-toothed cat, and a primitive horse and her foal. We can confidently attribute the humanlike tracks to *Australopithecus afarensis*, whose bones occur in nearby strata both below and above the ash. Two bipedal trackways are present, but one is double. The latter represents two individuals, one of whom walked across the soft ash in the footsteps of the other. The tracks in many ways resemble those of modern humans, although the big toe diverges more from the smaller toes.

More Recent Evidence of Part-time Climbing

By the early 1980s the evidence for bipedalism in *Australopithecus* had engendered a widespread belief that the creature was totally terrestrial—a confirmed ground dweller. This view was soon to be challenged, however. Further study revealed that, from its fingers to its toes, *Australopithecus* possessed features that would have made it a much more adept tree climber than an average modern human. In addition, by the standards of our species, it was no great walker or runner. In short, during the 1980s this ape-brained animal began to look like a creature that had not fully abandoned its ancestors' habit of climbing trees.

One piece of evidence that *Australopithecus* habitually climbed trees actually had surfaced in the 1970s but was temporarily ignored. This was the orientation of its shoulder socket. The upper limb bone of apes and humans terminates in a large knob that fits into a cuplike depression at the corner of the shoulder blade. The result is a ball-and-socket joint that allows for a wide range of arm movement. In humans the socket opens out to the side, allowing us to rotate each arm like a windmill. More importantly, it has allowed members of our species to hurl spears with an overhand motion and to carry objects comfortably at our sides and swing our arms easily when we walk or run. Apes have a very different shoulder configuration. Although we think of them as massive animals, their shoulders are actually quite narrow. This condition positions the shoulder socket close to an ape's center of gravity, which makes climbing easier. The socket in an ape's shoulder also faces diagonally upward. This orientation serves these frequent climbers well, allowing them to extend an arm overhead without tilting the torso.

It is not possible to reconstruct the shoulder breadth of *Australopithecus* from fossil remains, but we do have solid evidence that the shoulder socket was directed upward, as in apes. A shoulder blade is a fragile structure, seldom well preserved in the fossil record. The portion close to the socket is thickened for strength, however, and one fossil fragment of a shoulder blade of *Australopithecus africanus* that includes the socket is quite revealing. This specimen, labeled STS 7, represents a large male whose socket is angled upward, well outside the normal human range. The specimen shows that even large members of the extinct group had shoulders that were well configured for climbing.

Jack Stern and Randall Susman of the State University of New York at Stony Brook were the first researchers to offer a broadly based argument that *Australopithecus* habitually climbed trees. They noted that *Australopithecus* not only had apelike shoulder joints but also lengthy arms. Lucy's upper arm bone was about 85 percent as long as her thigh bone. In this ratio she was about halfway between a small chimpanzee and a human pygmy. Also serving *Australopithecus* well for climbing would have been its powerful wrists and hands. It had a large pisiform bone, like that

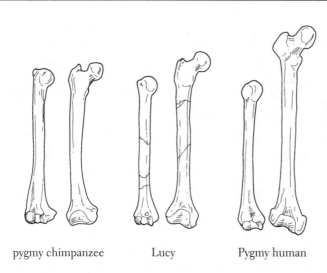

pygmy chimpanzee   Lucy   Pygmy human

FIGURE 2.2

Upper arm bones (left member of each pair) and thigh bones, showing that Lucy had long arms relative to her legs. In limb proportions Lucy was intermediate between humans and chimpanzees. (Thigh bone of Lucy reversed for comparison.)

of an ape. (The pisiform is the knoblike bone on the outside of the wrist against which a watchband rests.) A large pisiform increases the leverage of one of the three main muscles that flex the wrist. The fingers of *Australopithecus*, like those of apes, were also longer for its body size than are ours. Robust muscles obviously inserted into large depressions in the finger bones. Furthermore, these finger bones were curved, like those of an ape that conform nicely to tree trunks and branches. In short, from shoulders to fingers, the forelimbs of *Australopithecus* displayed apelike traits that would have made it a much better climber than any modern human.

Even the hind limb of *Australopithecus*, though adapted for walking and running on the ground, retained traits that would also have made it superior to modern humans as a climber. The toes were more curved than ours and longer for the length of the leg.

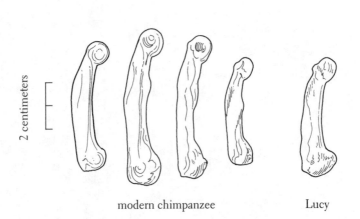

2 centimeters

modern chimpanzee   Lucy

FIGURE 2.3
A finger bone of Lucy (right), which was long and curved, like the finger bones
of a modern chimpanzee (left).

These, like the long, curved finger bones, would have been useful for curling around the branch of a tree or for gripping its trunk.
There is also new evidence that *Australopithecus* could oppose its
big toe to the other toes. With this kind of thumblike big toe, it
was truly quadrumanous, or "four-handed," like an ape. In 1995,
Ronald Clarke and Phillip Tobias of the University of the Witwatersrand reported the discovery of four key bones from the left foot
of a creature that was almost certainly *Australopithecus*. These
turned up in cave sediments from the Sterkfontein region of South
Africa that are at least 3 million, and probably close to 3.5 million,
years old. When assembled in their original configuration, the
bones reveal that even when the big toe was locked in place for
walking it did not point straight ahead but diverged from the other
toes. Furthermore, the form of a ball-and-socket joint deep within
the foot shows that the big toe was much more mobile than that
of a modern human. Clarke and Tobias concluded that the ani-

mal's foot differed by only a small degree from that of a chimpanzee.

In another feature of its lower leg that is likely related to a habit of vertical climbing, *Australopithecus* was truly remarkable. Detailed study of the creature's fossil ankle bones indicates that it was ahead of apes, to say nothing of humans, in its ability to flex the foot upward—to rotate the forefoot upward toward the shin. This flexibility would have been especially useful for climbing a tree with the feet applied to the trunk.

A Different Pattern of Climbing

Vertical climbing is now widely viewed as the habit that *Australopithecus* inherited from forebears who spent all of their time in trees. This ancestral habit has been inferred even from the anatomy of the human hind limb, which shares certain features with that of lower primates who are vertical climbers. The ancestors of *Australopithecus* presumably first employed bipedal locomotion in trees as a means of traversing horizontal branches while the forelimbs grasped higher branches for balance. Such animals would contrast with modern apes, which walk quadrumanously along limbs or swing from limb to limb. When more recent ancestors of *Australopithecus* descended from trees on all fours to take up a part-time existence on the ground, they made the transition on their hind legs.

A few researchers have argued that, although *Australopithecus* possessed many anatomical traits that would have made it better than a modern human at climbing, it did not climb trees as a matter of habit. These skeptics have pointed out that the extinct genus lacked the full range of bony structures that apes put to good use in their arboreal activities. This evidence does not rule out part-time climbing for *Australopithecus*, however. It simply shows that the creature did not match apes in its pattern of climbing or in its degree of mobility in trees. Apes, with their very long arms and grasping hind feet, are arboreal acrobats, swinging and leaping from branch to branch with great agility. *Australopithecus* presumably used its forelimbs primarily for ascending trees and for grasping branches as it walked horizontally along supportive limbs.

The pattern of vertical climbing that experts now envision for *Australopithecus* resembles the behavior of a telephone repairman who ascends a pole with a strap looped around both his waist and the pole and with cleats studding the soles of his shoes. The human climber quickly tosses the loop upward from a stable position and then takes a step upward with each foot. In the absence of cleats, *Australopithecus* would have used its long toes and acutely flexing ankles to grip the trunk of a tree. In place of a loop, it had long arms and powerful hands with big, curved fingers for grasping the sides of the trunk. The long loop allows the repairman's torso to tilt backward so that his weight strains against the loop, binding it to the pole. By having arms that were long in relation to its legs, *Australopithecus* gained the same kind of mechanical advantage. Its torso would have angled backward, giving its hands a firm grip on the sides of the tree while its feet stepped upward.

Even without the accoutrements of telephone repairmen, modern humans can climb in the manner just attributed to *Australopithecus*. Members of the Sakei and Semang tribes of Malaysia ascend palm trees so rapidly in this way that they are sometimes said to run up the trunks. Sailors of the same region even scamper up the vertical masts of ships by this method. These habitual climbers of the Far East are so experienced that they can actually splay their big toes so as to grasp a tree trunk or mast. In other words, the human big toe is somewhat prehensile, even if it cannot oppose any other toe, pad against pad. *Australopithecus*, with its long, curved digits, would have been even better at gripping trees with its feet.

Even in body weight, *Australopithecus* had an advantage over

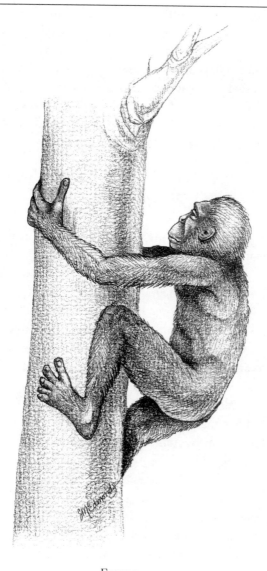

FIGURE 2.4

A reconstruction of *Australopithecus* in the act of upright climbing. The value of its long arms, fingers, and toes is evident.

modern humans as a climber. Beyond a certain size bigger is not better when it comes to arboreal activity. Modern humans are well above the optimum. The problem is that as an animal of any shape gets larger its weight increases more rapidly than its strength. In reality King Kong would probably have been too weak to stand up, let alone climb any structure big enough to hold him. On a smaller scale *Australopithecus* must have been a better climber than it would have been had it grown to our size.

A Life Divided

Taken at face value, the bodily configuration of *Australopithecus* would appear to represent an evolutionary compromise, combining both arboreal and terrestrial adaptations. Although modern apes are not good anatomical models of the ancient apes from which *Australopithecus* evolved, the ancestral apes shared with chimpanzees and orangutans some adaptations for climbing that, as we can see, they passed on to *Australopithecus*. These included upward-directed shoulder sockets, forelimbs that were long relative to the legs, strong hands and wrists, and long, curved fingers and toes. All of these features and also a relatively small body size would have made *Australopithecus* superior to modern humans as an upright tree climber. Even so, as I have already described, it was clearly less adept in trees than an ape. On the other hand, it was obviously less adept on the ground than a modern human. Let me review these terrestrial deficiencies.

The upward-directed shoulder socket of *Australopithecus*, while beneficial for climbing, would have been a hindrance to certain potentially useful terrestrial activities. Hurling a crude spear or even a stone as a lethal weapon was out of the question. By our standards even carrying objects at the side or swinging the arms while walking would probably have been awkward.

The relatively short legs that would have enhanced the ability of *Australopithecus* to climb trees inevitably reduced endurance in walking or running on the ground. The principle is quite simple. Had its legs been longer, the animal would have taken fewer strides per mile and therefore tired less quickly. Despite the greater stamina to be gained by evolving toward our long-legged condition, *Australopithecus* experienced no such change.

Along with its stubby legs, *Australopithecus* was equipped with feet that were longer than the optimum for long-distance walking. Lucy's forefoot, including her toes, was very long for the length of her leg. Mechanical analyses have shown that a relatively long forefoot causes the foot to swing slowly from step to step, reducing maximum walking speed. Lucy and her companions could not have kept pace with modern humans in power walking.

The long toes of *Australopithecus* would also have hindered it in running. Every time *Australopithecus* pounded the turf while on the run, it would have had to lift its body over the extended big toe, which would have remained in contact with the ground until the so-called toe-off occurred. The great length of the toe would have required that the animal's calf muscles apply a greater force to complete the toe-off than would have been needed with shorter toes. Quite simply, the long toes would have forced the animal to lift its body considerably upward before it could lunge forward to land on the other foot. Toes need be only long enough to provide balance, and the relatively short toes of modern humans are quite adequate in this regard. Our stable, forward-directed big toe also gives us a stronger grip on the ground in walking and running than the splayed, mobile big toe of *Australopithecus* would have provided him. What the long toes and mobile big toe of *Australopithecus* would have been useful for, of course, was climbing.

A waddling gait is the locomotory pattern that French anthropologist Christine Berge has attributed to *Australopithecus afarensis* from a biomechanical analysis of Lucy's pelvis and thigh bone. According to Berge, this gait entailed considerable rotation of the shoulders and pelvis about the backbone. Certainly, this was not the pattern of an adept runner.

In fact, *Australopithecus* may not have been able to run at all, except at a slow jog! Two Dutch scientists, Fred Spoor and Frans Zonneveld, working with the British researcher Bernard Wood have recently uncovered evidence to this effect that comes from, of all places, the creature's ear. Three bony, semicircular canals in the inner ear play an important role when humans balance on their toes or the balls of their feet in running or jumping. These canals form part of a system that unconsciously monitors upright movements. The relatively large anterior and posterior canals of humans have been found to be more sensitive than the smaller structures of apes. Three-dimensional X-ray imaging of fossil skulls of *Australopithecus*, including the Taung skull, reveal a canal system that remained stunted in the configuration of an ape. Was high gear for the extinct genus a shuffling lope or a trot? This creature certainly was neither nimble nor quick.

The body of evidence I have described portrays a creature that divided its time between two worlds. These worlds demanded different adaptations that were in evolutionary conflict—that got in each other's way within a single body. There was simply no way for *Australopithecus* to have become the equal of modern humans on the ground while matching an ape's prowess in the trees. Its retention of tree-climbing adaptations for a vast stretch of geologic time is powerful evidence that the animal did, in fact, spend a significant part of its life in trees. In other words, in locomotary abilities as well as in form, it was intermediate between apes and humans.

Whether one adheres to the gradualistic or punctuational model of evolution, it is impossible to deny that *Australopithecus* would have had a substantial opportunity to rid itself of its apish traits if it had not, in fact, been putting them to good use. Gradual evolution had more than a hundred thousand generations to mold this ancestor into something more closely resembling us, yet it accomplished next to nothing. Furthermore, evolutionary branching had multiple opportunities to produce punctuational steps in our direction but failed to do so. *Australopithecus africanus* diverged only slightly from *Australopithecus afarensis* when it branched off to become a distinct species; the body plan of the ancestral form remained essentially intact. A minimum of two additional branching events also took place in the evolution of *Paranthropus*. This heavy-jawed genus comprised three known species, at least two of which existed simultaneously. In limb structure *Paranthropus* differed only slightly from *Australopithecus*, and the origins of its species entailed relatively little change toward the human condition.

Confronting Evolutionary Stagnation

When I conceived of the arboreal imperative and the catastrophic scenario, I found it remarkable that anthropologists had ignored the general evolutionary stability of *Australopithecus*. This condition had not been generally acknowledged, let alone explained. This situation no doubt reflected the prevalence of gradualism at the time. As a proponent of the punctuational model of evolution, I was not about to overlook this salient example of evolutionary stagnation. Even so, in the 1970s, when *Australopithecus* was being portrayed as a full-time ground dweller, I had no explanation for its failure to evolve substantially in the direction of modern humans during a long stay on earth. On the other hand, once I had the arboreal imperative in mind, the explanation jumped out as soon as I became aware of the evidence that *Australopithecus* habitually climbed trees. The persistence of its tree-climbing traits was, of course, automatically explained by its spending part of its life in trees. What was more exciting, however, was that it was now apparent why it had failed to evolve the large brain of *Homo* for more than a million and a half years.

As explained in the previous chapter, Darwin believed that the great apes had not evolved brains like ours because, when these animals are on the ground, their limbs are so often occupied with quadrupedal walking and standing that they are seldom available for tool use. Only the development of upright posture and a tool-making culture, Darwin believed, would lead to the evolution of a brain like ours. It is now clear, however, that the evolution of bipedal posture on the ground did not trigger the evolution of a large brain. The two-legged posture of *Australopithecus* on the ground left the forelimbs free to manufacture, manipulate, and transport artifacts, including weapons. Still, in the course of a hundred thousand generations or so, *Australopithecus* failed to produce a stone culture and it failed to evolve a large brain. There must have been an additional barrier.

Here is where knowledge about the extreme immaturity of human infants comes into play—the immaturity that I described briefly in the previous chapter and will discuss more fully in chapter 5. The animals that evolved into *Homo* needed to have their hands *continuously* free to carry the helpless infants that would accompany an expanding brain. This required that the animals first abandon even their part-time arboreal activities. No mother could have climbed trees every day with an infant under one arm. This was the terrestrial imperative—the restriction that prevented *Australopithecus* from evolving in the direction of *Homo* until it was forced to take up a full-time life on the ground.

The World of Australopithecus

One might ask why the ancestors of *Australopithecus* ever began to descend from trees in the first place. Why did they begin to spend even *part* of their time on the ground? A fully arboreal existence had provided their predecessors with food and safety aloft, beyond the reach of most large predators. Did some sort of

environmental change alter this seemingly comfortable existence—a change that forced the upright climbers to test their balance on the ground? Here, although there was nothing for their
forelimbs to grasp for support, they could apply their feet to the
stable, broad surface that modern humans celebrate as terra firma.

Might previously dense forests have fragmented, so as to force
many populations of whatever creatures evolved into *Australopithecus* to make their way, awkwardly at first, from small grove to
small grove because their territories no longer contained large
woodlands in which they could move from tree to tree for an entire lifetime without running out of food? Exposure to ground-
dwelling predators would have been worth the risk if starvation was
inevitable in a small copse of trees. This, in fact, is the scenario
first suggested by Alfred Romer of Harvard and more fully developed by Peter Rodman and Henry McHenry of the University of
California at Davis. The change of behavior presumably took place
late in the Miocene epoch or perhaps as recently as five million
years ago, in the earliest Pliocene, when, genetic studies suggest,
the evolutionary lineage that led to humans branched from the one
that led to modern apes.

The problem is that fossil pollen reveal that vegetation in
Africa did not undergo any dramatic shift between about ten and
five million years ago. Forests remained so widespread that one
cannot look to their general shrinkage to explain the transition
from a totally arboreal creature to *Australopithecus*. It is possible
that a local population came under this kind of environmental
pressure, but, in truth, we have no idea why our ancestors began
to take up a part-time existence on the ground.

A localized environmental change did occur at the very end
of the Miocene epoch. The entire Mediterranean region, including the northern rim of Africa, passed through a crisis that affected
many of its inhabitants. In fact, it was through fieldwork in northern Italy that the British geologist Charles Lyell distinguished the
Miocene and Pliocene epochs of geological time. He founded the

epochs in a volume, published in 1833, of *Principles of Geology*, a
classic series that profoundly influenced Darwin on his epic voyage around the world aboard the HMS *Beagle*. Lyell laid the foundations for modern geology and filled in previously uncharted
segments of the geological timescale. Fossil seashells of the strata
that he distinguished as Pliocene differed from those of the strata
below, which he designated as Miocene. Today we recognize that
this faunal shift was a regional phenomenon that resulted from cataclysmic events in and around the Mediterranean Sea.

Whether from a northward lurch of the African plate or from
a slight drop in the level of the world's oceans, the Strait of Gibraltar somehow narrowed abruptly, slightly before six million years
ago. This constriction prevented the flow of waters from the Atlantic from keeping pace with evaporation, so that the Mediterranean shrank far below its present level. The narrow waters of the
Atlantic that crossed the threshold at Gibraltar must have cascaded to meet the declining surface of the Mediterranean by way
of a waterfall more spectacular by far than any that can be seen on
earth today. When the Mediterranean evaporated to its lowest
level, Atlantic waters plummeted more than a mile and a half before striking its surface. The evidence for this wonder of the latest
Miocene world came in part from studies that Soviet scientists undertook in preparing to build the Aswan High Dam across the Nile
River. The seismic analyses, or tracking of artificial earthquakes,
used to test the natural underpinnings for the proposed dam
yielded startling results. The seismic waves revealed a chasm,
larger than the Grand Canyon of Arizona, buried beneath the sediments of the modern Nile. The ancestral Nile had excavated the
canyon as it sliced its way downward to meet the declining surface
of the Mediterranean.

As the waters of the Mediterranean shrank, they became so
briny that salts crystallized from them and many forms of marine
life died out. Nearby land areas for which the Mediterranean had
previously supplied moisture and provided thermal stability be-

came seasonably arid and cool. In short, ecosystems in and around the Mediterranean were disrupted.

The Mediterranean crisis ended about a half million years after it had begun. At this time sea level rose, presumably because the Antarctic ice sheet melted back a bit, pouring water into the world's oceans. Around the globe shorelines retreated before advancing seas. The Strait of Gibraltar widened and deepened once again, and a strengthened cascade of Atlantic waters refilled the Mediterranean basin, restocking it with marine life. Some species never returned, however, and others arrived for the first time. The changed composition of the fauna was what led Lyell, who knew nothing of the vicissitudes described here, to recognize the post-crisis interval as the Pliocene epoch.

Although regions of Africa that bordered the Mediterranean experienced marked environmental change during the transition to the Pliocene epoch, *Australopithecus*—or its immediate ancestors—were probably little affected by the events along the northern margin of the continent. Presumably they continued to make a good living in the widespread woodlands far to the south.

A Chimplike Existence

What, then, was the behavior of *Australopithecus* in the patchwork of wooded and grassy terrains of Pliocene Africa? During the early 1980s, when anthropologists had reconstructed *Australopithecus* as a committed ground dweller, the animal invited ecological comparisons with primitive cultures of our own species. When, instead, we view this small-brained member of our family as a part-time tree climber, chimpanzees serve as a better, if imperfect, ecological model.

As I have previously noted, the menu of *Australopithecus* must have resembled that of chimps, in which fruits, seeds, and leaves dominate but insects and small mammals comprise a small but significant percentage of the caloric intake. The broader molars of *Australopithecus* nonetheless suggest greater emphasis on coarse fruits and seeds, which require crushing and grinding.

The foraging behavior of chimpanzees likewise serves as a reasonable model for *Australopithecus*, although presumably the extinct genus differed in the particulars of its feeding activity. Chimps feed mainly in trees. At Gombe National Park in Tanzania, where Jane Goodall has spent years studying chimps, they spend nearly half of their waking hours feeding and an hour or more a day moving between feeding sites. In other words, they exhaust local food supplies and then move on. Chimps normally forage in small parties that travel a total of two or three miles a day. They roam from home base to home base within territories that are large enough to supply food for the long term. For the most part, members of a troop forage in a small core area, visiting peripheral areas of their range only when food is scarce. This pattern typifies territorial mammals in general: they tend to maintain ranges that are large enough to support them during lean times.

Wherever they have their last meal of the day, chimpanzees build comfortable beds in the safety of trees. They weave branches into springy mattresses and often line them with pliant twigs and leaves. Members of a party colonize neighboring trees or sometimes crowd into one. When morning comes, they abandon their nests, although they may rebuild at the same site on a later occasion. *Australopithecus* may well have followed a similar sleeping ritual. Of course, in the manner of chimps, it presumably used trees as havens when predators threatened.

If *Australopithecus* resembled chimps in its general pattern of foraging and use of trees, it differed from them in some important aspects of its behavioral repertoire. It clearly was less skilled in arboreal acrobatics, having shorter arms and lacking the grasping big toe of an ape. Chimps range into the rain forest habitat, where they move gracefully through the lower canopy, fifty or even a hundred feet above the forest floor. It is chimps of more open and fragmented forests that serve as better ecological models for *Australo-pithecus*.

Though a less proficient climber than apes, *Australopithecus* may have been more efficient in locomotion on the ground. Tread-mill experiments have shown that chimpanzees expend much more energy than most other four-legged animals when walking or running at a given speed. They also expend much more energy than humans—over twice as much. Their problem is that the form of their limbs represents a compromise: apes must climb and also travel on the ground using both their forelimbs and their hind limbs. Their knuckle-walking style of locomotion on the ground is, in effect, an imperfect solution to the need to walk and run by using the grasping structures with which they climb trees. The hands of *Australopithecus* were spared this role in walking and run-ning. Although it was less well adapted than modern humans for movements on the ground, the hind limbs of *Australopithecus* benefited from a partial division of labor. Its heavy reliance on its forelimbs for climbing allowed the hind limbs to evolve special fea-tures, such as a springy arch resembling ours, for moving about on the ground. *Australopithecus* probably spent more time than chimps do in these terrestrial activities.

Judging from its slightly larger brain, *Australopithecus* was a bit smarter than a chimpanzee. Given this mental superiority and the freedom of its hands during bipedal locomotion, it almost cer-tainly made better use of simple tools than did any ape. Its ability to walk and grasp things at the same time would have allowed it to transport implements. It could also have carried food during its peregrinations. We probably will never know whether it employed containers to carry objects. One thing, however, is almost certain: it did not fashion stone tools on a large scale. Had such durable artifacts existed, we would have found them by now.

Even though *Australopithecus* resembled modern humans in having a long thumb, it appears to have lacked the ability to em-ploy many of the fifty or so different grips that we can put to good use in the aspects of our everyday activity that can loosely be termed "tool use." The small brain of *Australopithecus* presumably prevented the animal from evolving our level of dexterity: it simply remained too dim-witted to use the kinds of tools that require sophisticated manipulation. Living more like a chimpanzee than even a primitive population of humans, *Australopithecus* was without stone choppers or slicers or spear points. Here we appear to see the limitations of an orange-sized brain.

A life divided between climbing and ground dwelling placed *Australopithecus* in an evolutionary straitjacket. It could not move significantly toward the modern human condition in either brain size or capacity for locomotion on the ground. What I have not explained is why this creature continued to climb trees for more than a million and a half years. Nor have I explained exactly why this long interval came to an abrupt end. These critical issues are the subjects of the two chapters that follow.

CHAPTER THREE

Life Among the Lions

There is good reason to believe that natural enemies took a greater toll on populations of *Australopithecus* than did shortages of food. Quite simply, Africa offered an abundance of staples for omnivorous hominids, but a motley squadron of vicious predators patrolled its landscapes. *Australopithecus* and early *Homo* shared their world with several species of four-legged carnivores that could easily outrun these weak, weaponless hominids and overpower them with slashing tooth and claw.

We have good evidence that large predators frequently victimized *Australopithecus* throughout its long stay on earth. Part of this evidence comes from fossil bones collected from a cave deposit at the South African town of Sterkfontein. These remains suggest that large cats, and perhaps also hyenas, made a habit of feasting on *Australopithecus*. More general testimony comes from what is known of carnivores that roamed Africa in the time of *Australopithecus*. From the behavior of their living descendants, we can deduce that these animals would have preyed heavily upon our ancestors.

A Killer Ape?

Early in the twentieth century, some scientists actually believed that *Australopithecus* was more of a killer than a murder victim. It was Raymond Dart, the discoverer of the Taung skull, who most vividly portrayed *Australopithecus* as a bloodthirsty big-game hunter. At first he had viewed this extinct primate as a harmless omnivore that consumed insects, birds' eggs, grubs, and berries. He harbored other suspicions, however. The fossil baboon skull that had drawn him to the Taung site displayed a neat, circular hole near its top. Some years later, he reported that his thoughts had flashed to this baboon skull when the Taung child—the first known *Australopithecus* fossil—surfaced from the same cave deposit. Might the species of ape-man to which the Taung child belonged have bashed in this baboon's head?

Years later, Dart's suspicions became a guilty verdict. He convicted *Australopithecus* on the basis of a variety of broken bones found in caves of the Sterkfontein region. These fossils, he concluded, were the remains of *Australopithecus*'s victims. Between 1949 and 1965, Dart published a series of papers purporting to show that *Australopithecus* was a savage predator—a killer ape. He even claimed to discern a pattern of right-handed clubbing in the skulls of fossil baboons. Finding no sophisticated tools in the cave deposits with *Australopithecus*, he speculated that the animal's weapons were themselves mostly bones. In his biased eyes, fractured limb bones and antelope horns became daggers and large thigh bones, truncheons.

In his autobiography, written in 1959 with Dennis Craig, Dart recalled, "By 1929 . . . I was claiming that *Australopithecus* did not seek food and protection by climbing trees. He had hunted his food in the open and was a shell-cracking, bone-breaking, flesh-eating ape." Dart apparently assumed that *Australopithecus* was so formidable an adversary with its arsenal of bone tools that it had no need to find refuge in trees. He later seized on the 1947 discovery of a nearly complete pelvis of the extinct genus to bolster his views. Like the pelvic fragments of Lucy that were to be exhumed more than three decades later, this fossil pointed to upright posture, and Dart joined many other anthropologists in concluding that *Australopithecus* was a committed ground dweller. Adhering to the up-from-the-apes scenario, Dart assumed that human ancestors had come down from the trees on all fours and only later become habitual bipeds. He ignored the possibility, now widely favored, that we evolved from part-time climbers that were upright both in the trees and on the ground.

Dart even saw murder clues in the skulls of *Australopithecus africanus* itself. Any jaw of this species that was fractured or missing some teeth became a victim of internecine clubbing. Dart suggested that the harsh environment of South Africa had provided *Australopithecus africanus* with little plant food, driving it to carnivorism and even cannibalism. "*Australopithecus* lived a grim life," he wrote; "he ruthlessly killed fellow australopithecines and fed upon them as he would any other beast, young or old." Dart speculated that bipedal posture itself evolved as an adaptation for hunting, allowing *Australopithecus* to survey open country for potential prey from an upright stance on its hind limbs and to waylay them with weapons that it wielded with its freed forelimbs. For him, erect posture "was the first step toward humanity. The dominant factor . . . was our ancestors' finding that they could kill prey and protect themselves much more efficiently with the aid of a club held in their hands than with their teeth."

Dart's analyses of fossils were so wildly speculative that few scientists took seriously his claim that *Australopithecus* lived and died by wielding bony weapons. On the other hand, the 1950s and 1960s saw a sudden rise in popularity of the idea that hunting, from the time of the australopithecines, had infused aggression into the human psyche. In his book *A View to a Death in the Morning:*

Hunting and Nature in History, Matt Cartmill of Duke University addresses the history of the unfounded belief that humans are quintessentially ferocious hunters who inherited their aggressive tendencies from ancestral killer apes. Cartmill attributes the resurgence of this belief to several factors, including a loss of faith in human goodness that stemmed from the atrocities of World War II.

Australopithecus *as a Victim*

The focus on hunting *by* the human family continued into the 1970s, drawing attention away from the hunting *of* the human family by other animals. Taking exception to the prevailing attitude was Charles Brain of the Transvaal Museum. Brain's unassuming but sensible analysis of the South African cave deposits stands in sharp contrast to Dart's flashy speculations about three-million-year-old mayhem. Brain scrutinized the cave sediments and their fossils thoroughly and dispassionately and wrote the lengthy book *The Hunters or the Hunted?* In answering his title's question, he concluded that early members of the human genus, *Homo,* found in the younger South African cave deposits, were indeed flesh eaters who managed to evict four-legged carnivores from caves. *Homo* left behind not only the fragments of its own skeletons but also the bones of its victims, which bear the marks of cutting and chopping. Also present are implements of stone and bone that served for butchering. We will perhaps never know to what extent early *Homo* was a hunter and to what extent it was a scavenger, but it must have eaten plants as well as flesh, just as we do today. For *Australopithecus,* Brain's story was quite different. He found neither obvious tools accompanying its fossil bones nor any indications that this small primate consumed large animals. Instead, he found evidence that *Australopithecus* was one of the hunted.

Of special import to Brain's analysis was the group of beds in the Sterkfontein cave that are known collectively as Member 4. Brain determined the minimum number of individual animals that the fossils found there could represent. He did this by clustering the various bones and teeth of mammals in Member 4 into as few individual animals as possible. For example, if two skulls and three left upper leg bones of adult animals had been found for one species, then at least three (but possibly five) adults of that species were preserved. Following this procedure, Brain tallied a minimum of 348 individual animals. The collections contained no fewer than 45 representatives of *Australopithecus africanus*, several of them juveniles. The abundance of this species is just one aspect of a great preponderance of large primate fossils in Member 4; the fossil material also includes a minimum of 181 baboons, representing at least three species. Remains of the skulls and jaws of baboons and *Australopithecus* are especially numerous. A collecting bias has probably elevated the proportion of cranial fossils slightly (scientists in the field tend to set their eyes for them), but so few primate bones from below the neck have turned up that the great abundance of cranial material in collections must reflect a similar abundance in the rock.

Carnivores and scavengers commonly crack bones, but they tend to leave skulls and jaws lightly damaged. To Charles Brain and others, the concentration of skulls in Member 4 therefore identifies the total assemblage of fossilized bones as the residue of predators and scavengers. The implication is that either the large primates or the beasts that consumed them occupied the caves where the bones accumulated — or that both groups occupied the caves, perhaps incompatibly.

Most species of meat eaters in Africa today are not only active predators but also opportunistic scavengers that feed on carcasses they discover or pirate from other animals. Supporting the idea that such animals left the skeletal debris in the South African caves is the presence there of fossil bones that represent several horses (of an extinct species) and many antelopes (divided among about a doz.₋ᵔ species). *Australopithecus* and baboons might have frequented at least the outer portions of the caves, but horses and antelopes are rarely troglodytes. Meat eaters presumably dragged these hoofed animals, or their body parts, into the Sterkfontein cave that became their final resting place.

Elisabeth Vrba has added another piece of evidence. In the assemblage of fossil antelopes preserved in Member 4, juveniles are overrepresented; they are more abundant than would be expected if the antelopes had died from a variety of causes. Carnivores preferentially select young animals, along with the old and the sick, because they are easy marks. Carnivores might easily have left behind the array of bones in the cave.

A few of the Sterkfontein bones actually appear to bear the scratch marks of teeth. One fossil australopithecine skull is itself a smoking gun, pointing to a leopard as the killer of its owner. This skull was exhumed from a cave deposit at Swartkrans, about a mile from Sterkfontein. In both size and spacing, two neat circular dents in the surface of the skull perfectly match the tips of a leopard's projecting lower canine teeth. Donning the hat of a detective, Charles Brain attributed the dents to the bite of a leopard that was dragging its victim to safe storage, perhaps in a tree, which is where leopards of the modern world frequently stash their victims for several days of feeding.

The Taung child may have been the victim of an entirely different kind of predator: It may have died in the talons of a predatory bird. Lee Berger and Rob Clark of the University of Witwatersrand have noted that small fractures and puncture marks on the Taung skull match those that black eagles of present-day Africa leave on their prey. Dart believed that the Taung child was preserved in a cave that housed *Australopithecus*, but the burial site now appears to have been a depression beneath the nest of large birds of prey. Here bones of the birds' prey, including baboons and tortoises, accumulated.

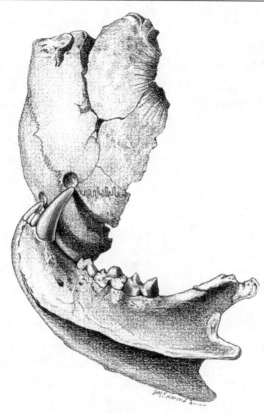

FIGURE 3.1

The teeth of the lower jaw of a leopard match pits in a fossil australopithecine skull from a cave at Swartkrans, South Africa. It appears that a leopard may have dragged the owner of the skull by its head.

The Real Killers

Where there are herbivores to be caught, there are generally carnivores to catch them, and trapped in Member 4 sediments at Sterkfontein are bones of some of the likely hunters of the primates and antelopes whose remains are also preserved there. Carnivorous mammals are much less abundant than their prey, both in life and also in the fossil record. This disparity is easy to understand. Every carnivore must consume many prey animals every year to stay alive, yet the prey are generally difficult to catch. As a result, many more zebras, gazelles, wildebeests, and other hoofed animals occupy Africa than do lions, hyenas, hunting dogs, cheetahs, and leopards.

Member 4 at Sterkfontein has yielded a few bones of five species of large predators, along with the remains of large primates and hoofed herbivores. Two of the fossilized predators belong to living species. One species is the spotted hyena, *Crocuta crocuta*, and the other is the modern leopard, *Panthera pardus*. Other fossils represent at least four animals of an extinct species of hyena that resembled a member of the dog family in having long, slender legs. There is also at least one true saber-toothed cat, *Megantereon gracile*, which brandished daggerlike canine teeth. Finally, there is at least one so-called false saber-toothed cat, *Dinofelis barlowii*, which resembled a leopard but was larger and stockier and equipped with more elongate, stabbing canines (see Figure 3.2). Just as "dinosaur," from the Latin, means "terrible lizard," *Dinofelis* means "terrible cat." Indeed, in single combat *Dinofelis* was undoubtedly more terrible than a leopard or spotted hyena. Thus, a formidable array of predators inhabited Africa along with *Australopithecus*.

At one time or another, the Sterkfontein cave may have served as a lair for each of the three species of large cats whose bones are preserved there. Modern leopards are known to lodge in overhangs and caves. Spotted hyenas generally dig dens for themselves in Africa today, but natural caves have sometimes served as their refuges in the past. Remarkable as it may seem, during the last glacial minimum, about 120,000 years ago, populations of spotted hyenas made their way to what is now Britain. There they left their dung and their bones in caves in which they apparently raised their cubs. The extinct carnivores preserved in Member 4 at Sterkfontein—the long-legged hyena and the true and false sabertooths—may have used the African cave in a similar way.

Modern Life on Open Terrain

So many of the predatory species that lived before the Ice Age have survived to the present that the activities of the modern fauna give a strong indication of the particular dangers that confronted *Australopithecus* more than two and a half million years ago. Open woodlands and savannas of modern Africa support a greater variety of large carnivores—and also more carnivores per acre—than do rain forests. The reason is easy to understand. The lush canopy of a rain forest so effectively shields the forest floor below from sunlight that little undergrowth is present and there is little forage for large, ground-dwelling herbivores. With few animals to hunt, carnivores are also relatively sparse. In contrast, the shorter trees and grasses of open woodlands and savannas offer abundant food for browsers and grazers. This is why vast herds of hoofed animals inhabit these regions and support a relatively high concentration of carnivores, including lions, wild dogs, hyenas, leopards, and cheetahs. *Australopithecus* spent its life among a similar concentration of carnivores, living as it did outside dense forests.

To understand the ecological role of large carnivores and the threat they posed to *Australopithecus*, we must begin one step below them in the food web, with the large herbivores that serve as their food. Tanzania's Serengeti National Park offers a splendid view of predator-prey relationships in the modern savanna-woodland ecosystem of East Africa.

The Serengeti Plain stretches from Lake Victoria eastward to the Eastern Rift Valley, just south of the equator. For many people the word "Serengeti" brings to mind a vast grassland, but the fact is that solitary trees and small groves dot the savanna. Most of these trees are acacias, whose parasol shapes are much photographed in profile when the sun is low. Few large herbivores feed on the taller acacia trees except for the leather-tongued giraffe, which can reach leaves far above the ground and can also tolerate the trees' nasty thorns. Savannas grade into woodlands, where trees are concentrated to the extent that their collective canopy, at a height of thirty or forty feet, typically shades 15 or 20 percent of the ground. Grasses occupy the open spaces between the copses of trees. Forests along rivers grow taller, sometimes forming canopies seventy or eighty feet above the ground. Savannas such as the Serengeti exist where rainfall is too sparse and too seasonal for woodlands to form. For nearly half the year, from late May to early November, drought desiccates the Serengeti and turns it brown. Then, in December, towering black clouds bring torrents of rain as the wet equatorial zone shifts southward. The Serengeti suddenly flushes with green and flowers bloom.

The savanna rejuvenates with greatest luxuriance where fires have swept over it during the dry season, returning nutrients to the soil from the dry grass and litter. In the modern world humans set most of these fires. Lightning ignites some, however, and lightning was the sole agent of ignition before our ancestors harnessed fire, long after *Australopithecus* had become extinct. Burning not only fertilizes the soil, but, by destroying small trees, it also prevents some areas of savanna from turning into woodlands. Grasses, unlike most small trees, regrow from their roots after a range fire.

Elephants also destroy woodlands by trampling, stripping, and uprooting saplings. So powerful is their effect that we can suppose that savannas would be more widespread today if populations of African elephants were restored to the levels that preceded the advent of modern human hunting.

About twenty species of large herbivores are common denizens of the Serengeti, but three species — the wildebeest, zebra, and Thomson's gazelle — greatly outnumber the rest. Near the end of May, members of these three dominant species join other large herbivores of the Serengeti in an immense migration northward. Remaining behind are rare species such as the hartebeest, sable, buffalo, and impala; these animals occupy sparse woodlands year-round. The more populous species of the broad grasslands must evacuate their browning habitat in order to sustain themselves. Wildebeests begin the trek in small groups, often walking single file. Like the tributaries of a river, the lines soon converge, eventually forming a vast herd of more than a million animals that flows toward the north. The wildebeests and other migrating species suffer heavy casualties during their long journey. Many drown in rivers that must be crossed along the way. The survivors spend most of the dry season in broad woodlands to the north, where food remains moderately abundant. Months later, when the herds return to the rejuvenated Serengeti, many of their members are scrawny. They quickly fatten on the newly sprouted grasses, however, and then, during a brief interval in January and February, females of many species, including the wildebeest, drop their calves or foals. This simultaneous birthing is no accident. It floods the savanna with infants. These ungainly newborns must be up and running within minutes after birth if they are to have more than a wisp of a chance for survival. Healthy infants are so numerous, however, that predators can manage to consume only a small fraction of them. If births were spread evenly throughout the year, few infants would escape predation.

The great migration that follows not only provides the herds with food during the dry season but also foils their enemies. By deserting the savanna with their young offspring, species of large herbivores leave Serengeti predators with few potential victims. A large percentage of lion and hyena cubs starve before the herds return, as do many of the pups of wild dogs.

Despite the seasonal exodus, the Serengeti is never without carnage. It is a place of deceptive calm. From time to time, the stillness is ruptured by sudden violence. Just as predators have evolved the ability to stalk and pursue, large herbivores have evolved the ability to flee. Large predators in Africa have no easy time capturing their prey. The top speeds of most species of predators and prey are roughly the same: thirty-five to forty miles per hour. The predators form two distinct groups with respect to strategy, however. Members of the cat family—the lion, leopard, and cheetah—are ambushers, built for rapid acceleration from a crouching start. Like the domestic cat, they stalk their prey and then dash a short distance, either making a kill quickly or abandoning the chase. Wild dogs and hyenas, on the other hand, are long-distance runners. Like wolves and domestic foxhounds of northern climates, they begin the chase without benefit of surprise and simply wear their quarry down, often after a mile or two of pursuit. From such behavior comes our adjective "dogged."

Tactics of Living Carnivores

The hunting behavior of carnivores on the modern Serengeti illustrates the dangers that *Australopithecus* would have faced had it attempted to live its life on the ground twenty-four hours a day. Consider first the largest of the predators, the king of beasts. Owing to the extensive observations of George Schaller of the University of Chicago, we have a detailed picture of the life of the Serengeti lion. Because of its size, the lion takes large species that seldom fall to other carnivores. One of the lion's preferred prey is the powerful African buffalo, which has lethal horns, and lions occasionally fell a giraffe or kill a crocodile or hippopotamus. Lions usually stalk in groups, spacing themselves at equal distances to form a linear phalanx that advances on their prey or a ring that closes in on them. Females, being lighter and quicker, do most of the hunting. Zebras, wildebeests, and Thomson's gazelles are their most common victims. Lions generally rush a group of herbivores after sneaking within ten to fifteen yards of them, and they pursue up to two hundred yards before abandoning a chase. If a lion hits one of the quarry, the other lions redirect their sprint and swarm over the victim. This method of attack yields success for lions 15 to 30 percent of the time. They do some hunting during the day, but much more at night, when they can more easily avoid detection.

For the same reason, they prefer the cover of woodlands to the openness of grasslands.

Prides of lions are highly socialized groups that, on average, include about ten animals. Because cooperation is necessary in the hunt, they seldom contain fewer than four animals. On the other hand, they seldom contain more than fifteen animals because a pride often manages to dispatch just one large herbivore during a single night's foray, a level of success that must feed the entire pride. For such a group in Pliocene times, a single carcass of *Australopithecus* would have amounted to no more than an appetizer. A small troop would have made a good meal.

On the Serengeti grasslands lions also derive 10 or 15 percent of their food from scavenging. They usually accomplish this by banishing smaller carnivores from a kill. They themselves are seldom driven from a carcass except by hyenas—and then only when badly outnumbered. A pride of lions maintains a territory large enough to support it even during the dry season. Lions do not generally migrate with the herds of their prey; not only are the small cubs ill-suited to long-distance travel, but so are the heavily built adults. Instead of migrating during the dry season, most prides shift from the grasslands to the woodlands of the Serengeti, where populations of herbivores, though relatively sparse, provide them with some food until the migrating herds return.

At the other end of the size spectrum of large Serengeti carnivores is the wild dog. Also known as the Cape hunting dog, this animal is a lithe, long-legged curser—highly adapted for pursuing prey over long distances. Erect, satellite-dish ears serve the wild dog well in detecting potential prey. Wild dogs live in packs of as many as forty animals unified by a complex web of social behavior. Following a kill, adult members of a pack return to a den dug deep into the savanna soil, and they regurgitate meat for pups that remain there. This behavior, and also the habit clan members have of licking mouths to greet one another, has had the tragic effect of spreading canine distemper, derived from domestic dogs, throughout the wild dog population of Africa. Today, wild dogs are on the decline.

Most frequently, wild dogs run down Thomson's gazelles on the Serengeti, and these victims are followed in number by wildebeests and zebras. The dogs make no effort to surprise their prey, but simply challenge them to a long-distance race. They tend to single out an especially vulnerable animal for pursuit—often one that is young, old, or sick. The wild dogs typically gallop forward along a broad front, and when the quarry angles to the left or right, the dogs on the appropriate flank take up the chase. The effect is to reduce the distance run to a nearly straight line, while the pursued animal takes a longer, zigzag path. Once engaged in a chase, wild dogs are successful slightly more than half of the time.

Hyenas are less appealing to humans than are wild dogs, but they suffer from an undeserved bad reputation. Their appearance is mangy and their vocal repertoire includes what to our ear are unattractive whoops and cackles, but they are more than the four-legged vultures they were once considered to be. Through extensive nighttime observations of the behavior of spotted hyenas, the Dutch behavioral ecologist Hans Kruuk found that they are not predominantly scavengers. They are actually skilled hunters that kill about two-thirds of the animals they consume on the Serengeti. They are also well socialized, though in an unorthodox way. The females, which are dominant, are slightly larger than the males and have a clitoris that is enlarged to the size of the male's penis. This unusual genital trait probably assists the females in maintaining their superior status, because when individuals meet, regardless of gender, they tend to engage in mutual sniffing and licking of genitalia.

Hyenas look superficially like dogs, but they are actually more closely related to the cat family and are still closer cousins of weasels. The spotted hyena is the most abundant carnivore of the Serengeti, ranging throughout Africa south of the Sahara except in the region of tropical rain forests. When I use the term "hyena,"

I am usually referring specifically to the spotted hyena. It is one of three modern species of hyenas in Africa, overlapping in South Africa with the brown hyena and in East Africa with the striped hyena, a species that also ranges into Asia. Kruuk spent about forty months studying the behavior of the spotted hyena in the 1960s, and his research has painted an entirely new picture of this animal, one of the predators whose remains were found with the great abundance of *Australopithecus* skulls at Sterkfontein.

Hyenas live in clans that range in size from two or three individuals to as many as eighty. A clan shares a den—a complex system of tunnels dug two or three feet below the surface of the ground. Hyenas sometimes hunt by night, but more often they scavenge for food during the day. Scavenging is easier in daylight because circling vultures mark the sites of kills, and a beeline course often yields an easy meal. After meat-slicing lions or wild dogs have abandoned a carcass, marrow-filled bones remain for hyenas to crack with their premolar teeth, which have evolved into huge, blunt cones operated by powerful jaws. Hyenas can devour and digest the entire skeleton of any carcass, and their dung is white with the calcium phosphate residue of bone.

When hyenas are not scavenging during the heat of the day, they are usually sleeping. At night, however, they become fearsome predators. Powerful shoulders and low-slung hindquarters give them an ungainly appearance, but they are world-class long-distance runners, and often pursue their nocturnal prey for a mile or two at speeds of up to forty miles per hour. On the Serengeti they apparently take wildebeests most frequently, followed by gazelles and zebras. They occasionally kill baby rhinos, but only at high risk of injury from incensed, horn-wielding mothers. From a large herd of wildebeests, hyenas will single out a particular animal to pursue—sometimes, it seems, by detecting aberrant behavior caused by illness or old age. When a wildebeest or gazelle is the quarry, one, two, or three hyenas form a hunting party for the high-speed chase. Zebras are trickier; a large group of hyenas—

sometimes as many as twenty—join in the chase, and an entire family of zebras, not just one animal, is the initial target. Zebra stallions are formidable opponents, however; they can inflict vicious bites with their grass-clipping buck teeth and have been seen to crush a hyena's skull with their flailing front hoofs. When a group of hyenas lights out after his family, a stallion drops back to fight a rearguard action to protect his harem of mares and their colts. About a third of the time he fails and, after a long chase, the hyenas swarm over some member of his family. The stallion is almost never taken.

Once they have their victim on the ground, the hyenas' enemies are the lion and, to a lesser extent, the wild dog. Only a large group of hyenas is likely to win a contest for a carcass against just three or four lions. A large pack of wild dogs usually defeats a much smaller group of hyenas in a contest over food. Although a hyena outweighs a wild dog by a factor of three, the quickness and cooperative attacking of the dog pack more often than not wins the day.

Unlike lions, wild dogs, and hyenas, leopards are solitary hunters. In fact, male and female leopards do not even live as pairs, but socialize only to mate. This isolated mode of life dictates many aspects of leopards' hunting behavior. They live exclusively in forests, woodlands, and thickets, where they can stalk to within a short distance of their prey to improve their chances of making a solo kill. Leopards also hunt mainly under cover of darkness. Although remarkably powerful for their size, they feed mainly upon impalas, gazelles, bushpigs, and other relatively small herbivores. They also take juvenile wildebeests and zebras, but adults of these species are too large for a leopard to bring down easily on its own. Only one charge in twenty by a leopard brings success. On the other hand, because of their habit of hauling carcasses into trees, these solitary animals lose very few of their kills to members of other species that hunt in groups, and a leopard—even one with cubs—can feed on a stored carcass for several days. Female leopards also protect their cubs by lodging them in trees.

Cheetahs, like leopards, usually hunt alone, but these swiftest of all animals inhabit the open savanna, where they run down small antelopes, especially Thomson's gazelles. Cheetahs resemble elongate leopards. With a shoulder height of thirty inches and an average weight just above a hundred pounds, a cheetah is slightly taller and lighter than a leopard. The physique of a cheetah sacrifices power for speed. Cheetahs are the greyhounds of the cat family. They stalk their prey in broad daylight without the need for protective cover. Once a cheetah launches an attack, it pursues its zigzagging target for up to four hundred yards at speeds of up to sixty miles per hour. Cheetahs seem almost too elegant to scavenge on the dead carcasses they encounter and, in fact, engage in almost none of this activity. With their small heads, blunt claws, and solitary habits, they are ill-equipped to contest other large carnivores for carcasses. High-speed chases are their métier.

Before setting out to hunt, females attempt to conceal their cubs in tall grass, but when lions discover cheetah cubs, they make a habit of killing them. This malicious behavior seems almost gratuitous, because lions compete hardly at all with cheetahs for food. Lions normally pass up small gazelles of the type that cheetahs favor in order to pursue larger animals that can supply an entire pride with a meal. Perhaps modern lions' penchant for murdering the offspring of another big cat traces back to a time when they experienced competition from the false sabertooth *Dinofelis*, whose bones occur with those of *Australopithecus africanus* at Sterkfontein.

The Enemies of Our Ancestors -

In the time of *Australopithecus*, the menagerie of large African predators resembled that of the modern world. If anything, it was more imposing. Already present were the modern species of lion, spotted hyena, brown hyena, and leopard. A cheetah was in existence and may have represented the living species, but its precise identity is uncertain. The hunting dog had apparently not yet evolved, but another species, now extinct, appears to have played much the same ecological role. This was the unusual long-legged hyena that I have already mentioned. It was obviously a long-distance runner and must have hunted in packs.

Dinofelis, which I have also described, was on the scene as well. It probably did some tree climbing, but it was more powerfully built than a leopard and was probably a less mobile climber. Charles Brain has noted that the long, stabbing upper canine teeth of this extinct false sabertooth would have served well for killing Australopithecus, and its powerful jaws and crushing rear teeth would have allowed it to consume every part of such an animal except the skull—the element most frequently left intact at Sterkfontein.

The fossil record shows that two distinct types of true sabertooths—cats with very long upper canine teeth—also occupied Africa with Australopithecus prior to the Ice Age. One of these, Megantereon, has been found in Member 4 at Sterkfontein. The other, Homotherium, has not. The canine teeth of Homotherium were slender, serrated blades that probably served to slice and tear the thick flesh of huge mammals. These fragile teeth would soon have broken against large bones if this sabertooth had habitually plunged them into antelopes, zebras, or members of the human family. In fact, the fossil record suggests that Homotherium was a specialized killer of elephants and perhaps other large pachyderms, an assortment of distantly related mammals whose name means "thick-skinned." The pachyderms include rhinoceroses and hippopotamuses as well as elephants, including mammoths. Homotherium ranged across a broad geographic region and survived until just a few thousand years ago. Some of its youngest fossil remains come from the Freisenbahn Cave of Texas. There about twenty adults and thirteen juveniles of Homotherium have been

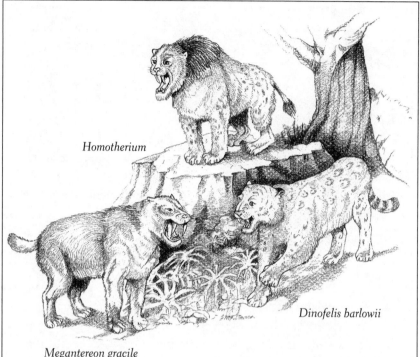

Homotherium

Dinofelis barlowii

Megantereon gracile

FIGURE 3.2

Two extinct species of large cats that occupied Africa at the time of *Australopithecus*, *Homotherium* and *Megantereon gracile*, were true sabertooths that probably used their bladelike canine teeth to stab elephants. On the other hand, *Dinofelis barlowii*, the "false sabertooth," was probably a threat to *Australopithecus*; it resembled a leopard but was stockier and had longer canine teeth.

recovered along with some seventy mammoths, nearly all of them juveniles belonging to a species that died out near the end of the Ice Age. It seems evident that this sabertooth tore through the thick hides of the young elephants with its serrated, saberlike teeth and then dragged the carcasses into the cave, where its hungry young were ensconced. This presumably was the general mode of operation of *Homotherium* in Africa as well, although the animal may not always have occupied caves. *Homotherium* was a long-legged animal that may have run down young elephants in open terrain.

Megantereon, the other true sabertooth of Africa, was shorter and stockier. Like *Dinofelis*, it probably occupied forests and woodlands and sometimes climbed trees, but its longer, daggerlike canines suggest that, like *Homotherium*, it stabbed pachyderms for a living. Even if the true sabertooths bypassed *Australopithecus* in favor of larger, less bony mammals, the group of predators that confronted this ancestor of *Homo* was more varied than the group that occupies Africa today.

It Could Not Hide

The predators that specialize in long-distance pursuit—hyenas and wild dogs—can sustain speeds of thirty to forty miles an hour for two miles or more, as can healthy zebras, wildebeests, and gazelles. Ambush predators—lions and leopards—sprint at even higher speeds. Natural selection has favored herbivores that are especially swift. A comparison of the running speeds of modern humans and African carnivores reveals just how vulnerable *Australopithecus* would have been on the ground. Large members of the cat, dog, and hyena families totally outclass even the fastest modern humans in sprinting and long-distance running. Very few humans can break the four-minute mile, but even those who can are managing only about fifteen miles an hour, far less than half the speed of a pack of wild dogs or hyenas. In fact, these four-legged coursers can run for a mile or two at nearly twice the speed of an Olympic sprinter!

Humans are poor runners for two reasons. First, our feet and ankles are not very springy. The reason a race horse must often be put to death after breaking a bone in its lower leg is that, because the fractured structure is spring-loaded, the tendons that keep it under tension tend to pull the separated fragments out of their proper positions. When a horse—or an antelope or a zebra—pounds its hoof against the turf, its ankle flexes at the fetlock, stretching elastic tendons that run along the back of the leg. When the stretched tendons then contract, they help to lift the animal as it bounds forward. Like the spring in a pogo stick, the tendons store energy and then release it again. Kangaroos make use of similar tendons in their hopping movements. We humans also have a spring system, but one that is much less efficient. Our arch, for example, acts as a spring. It flattens slightly when our foot strikes the ground, stretching a broad tendonous structure that runs most of the length of the foot. Our arch flexes only slightly, however, as do the Achilles tendon and other elastic structures of our foot and leg. We have to put much more muscular effort into each stride than do fast-running four-legged mammals, which, in effect, bounce from one pair of limbs to the other as they gallop.

Our second deficiency in long-distance running is that we have heavy legs that do not swing back and forth easily. Bipedalism is at the root of this problem. Because we have only two legs to support our bulk, they must be strong and heavy. Four-legged animals of our size have slenderer, more mobile legs.

Australopithecus was far less swift than even modern humans. Recall that the short legs of this animal saddled it with less endurance than we have because they required it to take more strides to cover a given distance. Recall also how the big feet of *Australopithecus* would have impeded the swinging of its legs and how its long toes, though useful for climbing, would have reduced both its speed and its endurance in running. Finally, bear in mind that this animal's apelike inner ear structures suggest that it lacked the balance to run as we do. It may only have jogged or shuffled along at a slow speed.

It Could Not Stand and Fight

Not only could *Australopithecus* not have outrun large flesh eaters, but it also lacked the ability to confront them in battle. Its natural weapons were meager indeed. It lacked even the dagger-like canine ("eye") teeth that groups of male baboons use on occasion to slash a leopard to death. In confrontations baboons frequently face down leopards, who recognize the danger and retreat.

Australopithecus lacked not only daggerlike canine teeth for self-defense but also stone-tipped weapons. Furthermore, its ape-like shoulder joints would have made for inept hurling of even simpler objects. Lacking our potential for windmill-like arm motion, apes generally throw objects with an underhanded toss. Like modern apes *Australopithecus* may occasionally have brandished sticks and stones, but with little effect against rapacious, group-hunting hyenas or lions.

Even baboons do not rely heavily on their canine teeth for defense, and the teeth did not evolve for this purpose. Sexual selection has produced the large canines in males; big males that have displayed the largest teeth have tended to intimidate others and monopolize the females. When it comes to dealing with predators, however, it is not the baring of teeth but flight into trees that usually sustains troops of baboons. These troops could not survive if they did not climb trees during the day when threatened or at night as a matter of habit. A habit of sleeping on the ground is also unthinkable for the defenseless *Australopithecus*. The most powerful predators of modern Africa—lions, hyenas, and leopards—engage in most of their hunting after dark, when they turn grasslands and woodlands into killing fields. Lions and hyenas are especially voracious at night because they hunt in groups that converge on their prey and then share the bounty. Because of their killing efficiency under cover of darkness, lions and hyenas are much more abundant than solo-hunting leopards. There is no more chilling sight in nature than a pride of lions silhouetted in the dusk, padding off in single file as they begin their nighttime hunt. Infrared cinematography shows how a pack of spotted hyenas, indolent scavengers during the day, also turn into a highly efficient killing force after dark. Lions and spotted hyenas were both on the scene in Africa when *Australopithecus* faced the shrinkage of its arboreal refuges.

During the 1970s and 1980s, when it had become customary to view *Australopithecus* as a full-time plains dweller, artists portrayed nuclear families of this feeble genus trekking across treeless savannas. In some of these ostensibly idyllic scenes, the heavy browridges of the subjects seem to express a well-justified apprehension. The artwork is often appealing in these reconstructions, but the science is untenable. An appropriate caption for them actually would be "When do the lions arrive?" Even modern Masai warriors, who have traditionally speared lions during the day, dare not spend nights in small groups on the open Serengeti Plain without the benefit of fire. *Australopithecus* lacked the spears of the Masai and had not harnessed fire. Hearths and remnant charcoal show that human ancestors did not begin to employ fire systematically until about a million years ago.

Peril from Fearless Carnivores

Humans' present relationship to undomesticated carnivores actually serves as a poor model for the interactions between *Australopithecus* and the predators with which it shared Africa. Just as we modern humans are afraid of large carnivores, they are wary of us. Their fear is learned behavior, perhaps passed from animal to animal by parental example or by one animal's observation of another's expression of fear upon encountering modern humans, who often tote guns. Large carnivores have learned these lessons well because modern humans are notoriously aggressive toward them. We have shot and trapped many of their populations into oblivion so as to preserve our domestic stocks or simply to mount imposing heads on our walls or flatten furry hides on our floors.

Clearly, a fear of humans is not built into the genetic architecture of lower animals. The earliest Europeans to settle in the New World were astonished not only by the abundance of the wild animals but also by their lack of alarm when humans approached. The native tribes were much less threatening than the gunpowder-laden newcomers soon proved to be. In fact, stories abound of the peaceable kingdoms that preceded the arrival of gunpowder in various parts of the world. The way things were is obvious to those of us who have been fortunate enough to set foot on the Galápagos Islands, where diverse finches and giant tortoises opened Charles Darwin's eyes to the reality of evolution. There, in accordance with regulations of the Ecuadorean government, visitors to many of the islands must travel on foot and in small numbers. The nesting booby birds, of three species, and the three-foot-long, cactus-eating iguanas take little notice of the visitors. For many generations of life, they have witnessed humans consistently confined to narrow footpaths that seem separated by invisible barriers from their positions just a few feet away. Humans look but do not touch—or shoot.

In Africa, where human behavior is less predictable, even the king of beasts bows to our presence. When humans are in the vicinity, lions hunt only at night. Despite their fear of us and our devastating weapons, however, lions continue to kill wayward or unguarded humans. In his 1972 book *The Seregenti Lion*, George Schaller reported that in 1961 a man died after a lion had dragged him from a tent by his head. Schaller also noted that at the time of his writing lions were beginning to add humans to their diet in the vicinity of Lake Manyara National Park. Such predatory outbreaks against rural African populations are quite common. In his book *The Spotted Hyena*, also published in 1972, Hans Kruuk painted a similar picture. Having developed great respect for hye-

nas during his three and a half years of field observations, Kruuk
sought to revamp their reputation. He noted that humans generally despise the hyena, which they view as a cackling, foul-smelling
scavenger to be persecuted at every opportunity. Kruuk nonetheless reported numerous attacks and systematic, repeated killings of
humans by hyenas in some regions of Africa during the 1960s. Yet
Schaller noted that hyenas often flee when a human on foot comes
within 1,000 feet—a greater distance than any other approaching
animal requires to put them to flight.

It is easy to imagine how much more aggressively large African
carnivores would have attacked *Australopithecus*, which lacked
threatening weapons. In their book *Cry of the Kalahari*, on desert
regions of southern Africa, Mark and Delia Owens reported on the
brazen behavior of lions, leopards, and hyenas that had never before encountered humans. As recently as 1974, these frightening
animals wandered into the Owenses' camp every night without
fear. Such unmitigated audacity on the part of a large carnivore
toward an early, weaponless species of the human family would
soon have led the carnivore to hunt that species with abandon.
Charles Brain, in concluding that leopards might have been major
contributors to the fossil assemblages of chewed bones of *Australopithecus* in South African caves, noted the tendency of these
big cats to turn into "man eaters" once they have tasted human
flesh. During recorded human history hundreds of leopards have
been known to turn into serial killers of humans in Africa or India.
Some have devoured hundreds of people.

The Need to Climb

If *Australopithecus* lacked effective natural or artificial
weapons and could not outrun African predators, how did it survive? Randall Susman and Jack Stern were the first anthropologists
to offer a cogent ecological argument that climbing should have

been an essential part of the behavioral repertoire of *Australo-pithecus*. Part of their argument pertained to diet and part to safety from predators. Many of the kinds of seedpods, nuts, and fruit pits that *Australopithecus* apparently chewed with its broad, thickly enameled teeth grew in trees. It is difficult to imagine why this animal should not have put its climbing prowess to use in gathering these dietary staples where they grew. Even so, feeding in the trees may not have been an absolute necessity on the basis of nutrition. More important than any limitation of food supplies on the ground should have been the threat of the large, voracious predators that prowled the African landscape. These marauders should have forced *Australopithecus* to climb trees as a matter of habit. No non-predatory species of prey that occupies open terrain and does not habitually burrow in the ground or climb trees can survive without having the ability to outrace carnivores. Because *Australo-pithecus* was not in this league of swift runners and could not have burrowed, it must have climbed.

The lives of large primates in Africa today offer powerful testimony about the dangers that *Australopithecus* faced on the ground. Leopards are the only large modern predators that can pursue climbing prey into trees, and since they are solitary hunters, they catch many fewer prey than do group-hunting lions and hyenas. Chimpanzees and baboons sleep in trees at night and flee into them during the day when predators threaten. Behavioral scientists have observed that when lions are present trees, as redoubts, are as important a resource for baboons as food and water.

Gorillas are the only large African primates that commonly sleep on the ground. They sometimes build low nests in trees, but often, in a remarkable display of residual instinct, they arrange flexible branches in a ring on the ground and then sleep within it. So ingrained in their genetic architecture is the habit of arboreal nest building that they somehow feel obliged to go through the motions, even though their effort serves no apparent purpose. True, gorillas do not have to sleep in the trees. Adapting the old joke, one might say that they sleep anywhere they want. They generally live outside the habitats of lions and hyenas, which would probably victimize them if they persisted in sleeping on the ground. Solitary leopards seem to leave gorillas alone.

Like troops of baboons or chimpanzees, bands of *Australopithecus* must have put their considerable climbing skills to good use both at night and during the day. At dusk they must normally have ascended trees for safe sleeping. During the day a head start may frequently have allowed them to beat much swifter predators in a handicap race to the nearest tree. Arboreal refuges, then, must have sustained at least a hundred thousand successive generations of *Australopithecus*.

What happened to bring an end to this long-lived lineage? As the next chapter will reveal, the world turned against our venerable ancestor, throwing it to the lions.

CHAPTER FOUR

When Winters Began

The Ice Age that shattered the world of *Australopithecus* about two and a half million years ago—what is called the modern Ice Age—was by no means the first glacial interval in the history of the planet. Earlier glacial episodes had already transformed life on earth numerous times. The earliest yet discovered took place more than two billion years ago, when continental glaciers spread over what is now North America. These rivers of ice plowed up gravelly debris, and when they melted, they abandoned coarse glacial sediments, known as glacial till, as evidence of their movements. Icebergs that broke from these glaciers along the shorelines of ancient lakes drifted away from the land, carrying cobblestone and boulders that fell from the melting ice and settled through the water and into the mud below. These so-called dropstones now sit conspicuously in the layered rocks formed from the mud, giving mute testimony to the existence of glacial ice when the earth was only half its present age. At that time only bacteria populated the world, and we have no idea what effect the glacial interval had on these primitive forms of life.

Death by Refrigeration

Cold temperatures that accompanied more recent continental glaciations clearly damaged life on earth. Some 450 million years ago, for example, when a huge continent slid over the South Pole and accumulated vast ice sheets, biotic repercussions extended thousands of miles to the north. This southern landmass contained the modern continent of Africa, where, in the region of the modern Sahara, glaciers left widespread tills that have long since hardened into solid rock. At the time of this glacial episode, advanced forms of life had not yet made their way onto the land, so that its victims were confined to the seas. As the seas cooled, many marine animals, including groups of trilobites and primitive mollusks, died out in one of the greatest global extinctions in the history of the planet. Thus ended the Ordovician period of earth history.

At the time Africa, South America, Australia, Antarctica, and the Indian peninsula were united to form a vast supercontinent in the Southern Hemisphere. Today we can treat these far-flung continents as pieces of a jigsaw puzzle and neatly reassemble their outlines into an image of the composite landmass, which has been named Gondwanaland, after a region of India. Throughout most of its lifetime, which spanned half a billion years, Gondwanaland wandered about the southern half of the globe. In its later years, from the Coal Age until midway through the Age of Dinosaurs, it temporarily attached to the southeastern margin of the United States. The collision that welded the two continents together thrust up the southern Appalachian Mountains. So large was the landmass formed by this union that, after just a bit of southward movement, it encroached on the South Pole and glaciers expanded again. They appear to have reached their greatest extent just before the beginning of the Coal Age, formally known as the Pennsylvanian period. At this time, nearly three hundred million years ago, marine life suffered another major extinction.

Much later in geologic time, just 34 million years ago, glaciers expanded near the South Pole yet again, and extinction swept away many forms of life. This happened at the end of the Eocene epoch, halfway through the Age of Mammals, when continents and oceans were approaching their present configuration and plants that belonged to modern families cloaked the land. Fossils of these plants, which geologists employ as thermometers for the distant past, tell a story of widespread environmental changes at this time. Because they left a strong imprint on the modern world, these changes have special importance to us. In many regions climates grew cooler and drier, on the average, and contrasts between summer and winter became more pronounced. The fossilized leaves of ancient trees provide key evidence for this. Plants whose leaves have jagged or toothed margins prevail in temperate zones, whereas smooth margins are the rule for the tropics. Late in Eocene time, the percentage of fossil species with smooth margins declined in many regions, and the percentage with jagged or toothed margins increased correspondingly. Plants adapted to dry conditions proliferated along with those adapted to cooler conditions.

The global change that transformed floras about 34 million years ago issued from the southern end of the earth, as it had at the end of the Ordovician period and the start of the Pennsylvanian period—and once again the cause was continental movement. After the initial breakup of Gondwanaland early in the Age of Dinosaurs, three fragments of the southern landmass—Africa, India, and South America—had headed northward. The rest of the supercontinent, which included the modern continents of Antarctica and Australia, had remained at the southern end of the earth. Finally, about 34 million years ago, Australia parted company with Antarctica, which was then left isolated over the South Pole, where it still sits today. The earth's rotation then caused a great ocean current to encircle Antarctica and, in the absence of direct sunlight, become quite cold. Currents that swept close to this so-called Circumpolar Current from farther north became partly trapped in it, and water that cycled round and round in the gyre grew colder and colder. No longer warmed by the ocean currents from lower latitudes that once bathed its margins, Antarctica cooled down. Under these conditions snow failed to melt away in the summertime. Instead, it piled up and compacted into glacial ice.

The rifting of continents that left Antarctica alone at the end of the earth set in motion the refrigeration system that today keeps the deep sea close to freezing throughout the world. Then, as now, cold, dense water that became trapped in the current encircling Antarctica sank to the bottom of the ocean, where it spread even to high northern latitudes. Groups of floating microscopic marine organisms that were favored by frigid conditions began to flourish in a belt around Antarctica. When they died, their tiny skeletons sank to the floor of the deep sea and were buried in mud. Today, scientists sample these telltale remains by lowering long tubes from ships in order to extract plugs of sediment from the seafloor.

Some of the cold water that descended to the deep ocean around Antarctica and spread northward welled up along the margins of continents at low latitudes, replacing warmer surface waters that winds dragged away. The cool surface waters that originated in this way chilled air masses in many parts of the world. They also supplied the atmosphere with less moisture, by way of evaporation, than warm waters had contributed before the upwelling of cold waters began. Colder winters and drier summers were the result. Many forms of life found this pronounced seasonality intolerable, and widespread extinctions transformed the global ecosystem. In many regions only hardy species remained.

Forty Million Years Ago—a Balmy World

This harsh new world was, in fact, our kind of world—with the kind of climate that today inflicts severe droughts on California and hard freezes on central Florida. It stood in sharp contrast to the world of the Eocene before Antarctica froze up. During the earlier time of global warmth, about forty million years ago, palm trees had flourished in a moist, subtropical Wyoming, and strong winds occasionally blew the fan-shaped fronds of these trees into a large lake in the Green River basin, where they became entombed in fine-grained sediment. A massive slab of Green River rock that displays one of these huge botanical specimens now adorns a wall of the Smithsonian Institution's National Museum of Natural History. This was a lake that never froze in winter. Among its inhabitants were crocodiles, whose cold-blooded metabolism requires year-round warmth of the sort now found in the United States only in the Deep South. Fossils also tell of balmy Eocene temperatures across the Atlantic. Remarkable as it may seem, southern England supported forests that resembled the jungles of modern Malaysia; more than half of the families of plants that lived there are restricted to the tropics in the modern world. Today, fossil seeds of tropical plants wash from the London Clay where it borders the mouth of the River Thames. At the end of the Eocene, the flora that shed these seeds gave way to a community of plants adapted to much cooler conditions. Needless to say, since this change England has never been the same.

It is possible to view the onset of the modern Ice Age, when *Australopithecus* was still roaming Africa, as an accentuation of the global climatic shift that brought the warm Eocene epoch to a close. The modern Ice Age, however, was centered not in the south, where the cooling had begun, but in the far north, where glaciers spread into Europe and the United States. For more than thirty million years after glaciers formed on Antarctica, the Arctic Circle of the Northern Hemisphere had remained largely free of ice, even though its climate was cool.

Many regions of the world not only became cooler 34 million years ago but also progressively more arid. Summers, especially, became drier. Forests shrank and were replaced by open woodlands and savannas, in which solitary trees and small groves stood far apart and tall grasses carpeted the spaces in between. In general, grasses flourish under drier conditions than those required by the kinds of trees that form dense forests. The general drying trend of the past 34 million years is evident not only in the fossil record of pollen but also in the fossil record of mammals' teeth. Grasses are harsher fodder than most leaves; they contain tiny bits of silica that discourage grazing animals. The result is that those mammals, such as modern horses, that habitually feed on grasses have tall, continuously growing molar teeth with convoluted surfaces. (Even these special feeding tools, however, wear down in the course of time, and this is why, in defiance of the adage, one can actually gauge the age of a gift horse by looking it in the mouth.) Species with such built-in grinding mills became progressively more common as the Age of Mammals progressed toward the Ice Age. Species that could not feed on harsh grasses, but only browse on the soft leaves and twigs of trees in the manner of a modern deer, were restricted to the shrinking forests and became correspondingly rarer.

Global climatic change was responsible for the general spread of grassy environments and the faunas adapted to them, but the uplifting of mountains played an important role in some regions. For example, elevation of the Sierra Nevada, about four million years ago, left the Great Basin in its rain shadow; moist winds from the Pacific Ocean rose along the western slope of the new mountain range and cooled, quickly dropping their moisture as rain. Evergreen forests that had cloaked Nevada, to the east, then died out, as arid conditions resembling those of the present spread across the Great Basin.

The cooling and drying of climates during the past few million years has by no means affected all regions of the world, nor has it followed a simple trend. Ironically, just before the modern Ice Age began, during the Pliocene epoch, a brief pulse of warming spread throughout many regions. Perhaps some natural process expelled an excessive dose of carbon dioxide into the atmosphere, intensifying the greenhouse effect in the way that the burning of fossil fuels is doing today. In any event, the Arctic region became warm—perhaps as warm as it had been since the Eocene epoch. The fossil record reveals that the Arctic Ocean must have been free of ice, at least in summer; more than a hundred species of mollusks adapted to temperate waters migrated through its waters from the Pacific to the Atlantic. Among these pilgrims was a species that is now conspicuous in fish markets: the blue mussel, which blankets rocky shores of Europe and North America.

The Pliocene warmth that preceded the modern Ice Age was truly global. It shrank the ice cap perched on Antarctica, releasing meltwater that raised the ocean higher than it stands today. As a result shallow seas flooded well inland of the present coast of eastern North America. In Virginia, for example, the Atlantic shoreline lay some seventy-five miles to the west of the modern seacoast resort of Virginia Beach. Fossil marine life preserved in this region points to a brief interval of mild, subtropical conditions similar to those of southern California today. To the south, the elevated seas inundated all of peninsular Florida, save a central upland, which stood high and dry as a small, circular island. During this warm interval tropical species that are now confined to the region of Miami and the Florida Keys ranged halfway up the Floridian peninsula, and coral reefs extended well north of the shallow waters off the Florida Keys, to which they are confined today.

Europe also enjoyed mild conditions just before the modern Ice Age. A warm arm of the Adriatic Sea inundated Italy south of the Alps, and fossil seashells reveal that the Mediterranean was essentially tropical at this time. Fossil leaves and pollen point to subtropical conditions on the land as far north as Poland. In many regions climates were also wetter than today, probably because the warm seas that bordered them supplied abundant moisture. Cypress swamps fringed the northwestern Mediterranean, for example, where the land is now much drier. *Australopithecus* gamboled about to the south of the eastern end of the Mediterranean in warm, forested terrain that is now desert and savanna.

The Modern Ice Age

What I have described for the Pliocene epoch was the calm before the storm. About two and a half million years ago, glaciers advanced across continents of the Northern Hemisphere. In Iceland shells of the mollusks that migrated across a temperate Arctic Ocean during the warm interval that preceded the Ice Age became entombed in strata slightly more than three million years old. Then glaciers spread into the region, depositing gravelly tills right on top of the fossil-bearing strata. Suddenly, the Ice Age was under way.

Since slightly before three million years ago, ice sheets have waxed and waned in the Northern Hemisphere but have never totally wasted away. Half a million years later, a few pulses of glacial growth produced successively larger glaciers, and lobes of ice reached the shores of the Atlantic Ocean, where they launched icebergs. Cores of sediment extracted from the floor of the central North Atlantic tell the story. They bear conspicuous grains of sand in layers that are about two and a half million years old. Normally only fine clay, much of it windblown, finds its way from the land to the deep sea. The oldest grains of sand in sediments cored from the floor of the North Atlantic mark the earliest arrival of icebergs. Some of these carried coarse particles of sediment as they melted.

With the expansion of numerous lobes of glacial ice to the seacoast, the young Ice Age crossed a developmental threshold about two and a half million years ago, just before big-brained *Homo* evolved and began to invent tools. At this time episodic glacial expansions, or glacial maxima, as they are called, began to attain great magnitudes. During each of these glacial maxima, ice sheets grew from centers in Scandinavia, Greenland, and Canada to cover broad areas of Europe and North America.

Radiocarbon dating of glacial debris provides a precise chronology for the most recent pulse of glacial expansion, which peaked only about 20,000 years ago. Ice sheets then retreated rapidly between about 12,000 and 7,000 years ago. Today, although the earth is approaching a glacial minimum, the Ice Age continues. The next glacial maximum will come slightly less than 100,000 years from now, unless humans burn enough fossil fuel to ward it off with greenhouse warming.

How is it possible to predict the glacial future in the absence of an artificial greenhouse effect? The answer lies not in the stars but in the earth's own movements. In recent years geologists have shown that glacial maxima have occurred with remarkable regularity. During the past nine hundred thousand years or so, glacial maxima have been separated by intervals of about ninety thousand years. Both the regularity of the maxima and the lengths of the intervening intervals require explanation. The answers seems to lie in the regular oscillations that characterize the earth's pattern of the rotation about its own axis and about the sun.

The orbit of the planet is not circular but elliptical, and its shape lengthens and shortens with a periodicity that varies slightly through time but averages about 95,000 years. In other words, the orbit is "out of round" in the extreme this often, and the planet then receives relatively little sunlight at the time of year when it reaches the outermost part of its orbit. This regular pattern appears to explain the expansion of glaciers every 90,000 or 100,000 years during the last 900,000 years.

Earlier in the Ice Age, a regular change in the angle of the earth's axis of rotation seems instead to have governed the waxing and waning of glaciers. Every 41,000 years, this angle is farthest from vertical, which means that the north and south polar regions are aimed farthest away from the sun during their respective winters—and winters in both hemispheres are especially cold. During these intervals of especially cold winters, glaciers expanded to their maximum extent.

It is not known why the Ice Age has had two phases, an early phase in which the varying axial tilt of the earth seems to have controlled the expansion and contraction of glaciers and a later phase in which the changing shape of the earth's orbit seems to have dominated. Nonetheless, in tandem, these factors seem to explain the glacial maxima and minima.

The Chronology of Fluctuations

How is it possible to date the maxima and minima—how do we know that they have occurred so regularly? Few glacial tills are amenable to dating, but sediments far removed from glaciers—those of the deep sea—have come to the rescue. In particular, atoms of oxygen in fossils preserved in the deep sea provide a record of the expansion and contraction of glaciers on the land.

Oxygen comes in two varieties, or isotopes: oxygen 16 and oxygen 18. Oxygen 18 has two more protons than oxygen 16. Although the protons are chemically inert, they add weight to the oxygen atom. The creatures that provide the most useful fossil record of oxygen isotopes are foraminifers, which are single-celled amoeba-like protozoans, some of which float in the ocean. Forams, as they are nicknamed, incorporate oxygen in the tiny skeletons of calcium carbonate they secrete. Atoms and molecules in fluids are always in motion. Molecules that contain oxygen 18, being heavier, move more sluggishly in water than do molecules that contain oxygen 16. The relative mobility of the two isotopes varies with temperature, however. The result is that less oxygen 18 ends up in the skeletons of floating forams at high temperatures than at cold temperatures.

In the course of millions of years, as forams have died, their skeletons have rained down on the floor of the deep sea, leaving a fossil record in the sediment that accumulates there. Through time this record reveals a zigzag pattern for the ratio of oxygen isotopes in foram skeletons. The proportion of oxygen 18 has been highest during glacial maxima, when waters have been coolest, and lowest during glacial minima, when they have been warmest. Of course, temperature changes have not been the same everywhere, so that the zigzag pattern differs somewhat from place to place.

A second factor amplifies the oxygen isotope signal. As it turns out, the entire ocean becomes enriched in the heavier isotope, oxygen 18, during glacial maxima. This happens because oxygen 16 tends to get locked up in glacial ice. The reason is easy to understand. Molecules of water—H_2O—that contain oxygen 16, being lighter than ones containing oxygen 18, evaporate more readily from the ocean. The result is that a disproportionate amount of oxygen 16 ends up as water vapor in the atmosphere. When glaciers expand by trapping snow, they automatically sequester water from the atmosphere that is rich in oxygen 16, and the ocean is left with a relatively large proportion of oxygen 18. Thus, the entire ocean shifts in the direction of this heavier isotope.

In other words, when glaciers have expanded, oxygen 18 has increased in the skeletons of floating forams for two reasons. First, under the cooler conditions, the forams have incorporated more of this heavier isotope. Second, with the expansion of glaciers, there has been a higher percentage of oxygen 18 in seawater to begin with. This double effect has endowed the fossil record with a powerful signal of the waxing and waning of glaciers. By measuring the signal in cores of sediment taken from the deep sea, paleontologists take the pulse of the Ice Age.

Early Skepticism

During much of the nineteenth century, the idea that there was an Ice Age at all met with a reception as chilly as the event itself. In fact, the proposition that glaciers had spanned continents not long ago, geologically speaking, led to one of the most vitriolic and protracted disputes in the history of geology—probably rivaled only by the twentieth-century debate over continental drift. In each case the radical idea eventually won the day.

Early in the nineteenth century, the idea of continental glaciation at midlatitudes offended naturalists not so much because it seemed unnatural but because it seemed unbiblical. Most geologists viewed the sequence of rocks on the earth as a record of catastrophic deluges, the last one being the biblical flood that Noah was supposed to have survived. Ice sheets had no place in this worldview, which was founded on religious faith rather than scientific observation.

Many facts that might have swayed opinion in favor of a recent ice age had yet to be recorded. The one great ice sheet of the Northern Hemisphere that has persisted during the present glacial minimum is the Greenland ice cap, but it was not until 1852 that a scientific expedition revealed that glaciers visible today near the seacoast of Greenland actually coalesce to form a single ice sheet covering most of the continent.

Much of the early controversy about glaciation focused on glacial erratics—large boulders that differ in composition from the bedrock on which they perch and that clearly originated from other bedrock positioned tens or hundreds of miles away. The biblically inspired interpretation had the great deluge rolling the erratics down mountain valleys and onto gentle plains.

Even scientists who had abandoned religious explanations in favor of everyday processes as agents of geological change found it difficult to imagine that veritable mountains of ice had once plowed over regions that farms and towns currently occupied. These skeptics found refuge in another hypothesis that invoked ice in a less radical way. Their alternative for the transport of erratic boulders and bouldery till was actually a kind of hybrid between the glacial hypothesis and the flood hypothesis. They envisioned an outbreak of icebergs that floated the rock debris to its present locations. This idea found favor with several eminent scientists, including Charles Darwin until late in his life, and its general popularity supplied the lexicon of geology with the term "drift," which is still widely applied to glacial till even though we know that it is no longer apt because glaciers do not float till into place, but push or carry it along.

Quite early in the nineteenth century, untutored but perceptive Swiss peasants actually surmised that glaciers had pushed boulders down Alpine valleys and into the plains where they now sit. A few scientists went further, adopting the idea of widespread glaciation. Among these was the Scotsman James Hutton, who is widely regarded as the father of modern geology because of his recognition, shortly before 1800, that natural processes such as the deposition of sediment by moving water and the eruption of lava by volcanoes produced the rocks we see all around us.

It was Louis Agassiz, however, who became the most influential champion of the glacial hypothesis. The Swiss-born Agassiz, who spent most of his career at Harvard University, was a distinguished paleontologist noted for his studies of fossil fishes. In an address to the Swiss Society of Natural Sciences in 1837, he aired the radical proposal that glacial ice had not simply flowed to lowlands from Alpine valleys in the recent geological past but had covered most of Europe and much of North America. In other words, from evidence of widespread deposition of materials by ice, Agassiz envisioned a true Ice Age.

Agassiz was on target, of course, but he went too far, leaping to an unjustified conclusion that impugned his credibility. He claimed that the big freeze should have destroyed all life on earth. Quite obvious here is the influence of Agassiz's mentor, the eminent French paleontologist Georges Cuvier. Cuvier, the reigning catastrophist of his day, contended that a series of violent cataclysms of unspecified origin had repeatedly denuded the planet of all plants and animals. In fact, Cuvier made a case for the reality of extinction in an era when a species' disappearance—one might say its ultimate failure—seemed at odds with the prevailing concept of a perfect divine creation of life on earth. Cuvier contended that no species as large as the European mammoth, whose skeletons were conspicuous in the fossil record, would have escaped notice if it walked the earth today.

Agassiz went further, citing the mammoth as a particular victim of the Ice Age. During the seventeenth century, Europeans had already learned that mammoths had been preserved in what amounted to natural frozen-food lockers. Some of these hairy elephants remain ensconced in Siberian ice that persists from the latest glacial maximum and yield meat that is still palatable to wolves and sled dogs. These frozen pachyderms apparently fell to their death in crevasses that cleaved glaciers, but Agassiz saw them as victims of the sudden freeze-up that he believed had engulfed the entire planet.

Agassiz's excessive claims worked against the general acceptance of the glacial hypothesis. The geologists who rose to prominence toward the middle of the nineteenth century, led by Charles Lyell, believed that natural processes had molded the earth and its biota, wreaking their changes slowly but powerfully, like an enormous machine in low gear. They were generally correct, but they failed to give abrupt natural disasters their due. Along with catastrophism in general, they rejected any concept of a glacial age. For a time Darwin joined otherwise progressive geologists, including Lyell, in clinging to the venerable drift theory. The drifters could point to large ice floes in modern seas of the far north—floes that seemed capable of invading continents if the sea level rose dramatically. Continental ice sheets, on the other hand, seemed unnatural until scientists took measure of the Greenland ice cap in 1852.

By the middle part of the century, however, two observations threatened the drift hypothesis. One was the absence of marine fossils from nearly all glacial debris. If flood waters had carried the gravelly debris in sea ice, they should also have left behind vestiges of animals that inhabited the rising ocean, but none could be found. The other problem was that some of the debris lay fully a mile above modern sea level, at elevations that seemed far beyond the reach of a flooding ocean.

Triumph of the Glacial Hypothesis

With serious problems having arisen for the drift hypothesis, evidence soon weighed in on the other side of the balance, favoring continental glaciers as the bringers of till and erratic boulders. In 1875 the Swedish geologist Otto Martin Torell created a great stir by reporting that scratches and polished grooves on limestones near Berlin mimicked features that small mountain glaciers had ground into rocks in Scandinavia. This was compelling evidence that glaciers had spread southward from Scandinavia across the Baltic Sea and into the lowlands of central Europe.

The likelihood that glaciers had scoured bedrock in Germany triggered a flurry of research on the gravelly deposits that might be attributed to continental ice sheets. It was soon apparent that, here and there, soils formed during periods of warmth lay between layers of this coarse debris. Intervals of glacial expansion obviously had alternated with warmer intervals. By the final decade of the century, geologists not only accepted the reality of the Ice Age, but they also saw that it was a complex interval of glacial expansions and contractions. The French word "moraine," for glacial deposits, emerged in the literature late in the eighteenth century, long after peasants had begun to apply it informally to gravelly ridges.

By the 1870s geologists had learned that a ridge of glacial till spans the United States, in places rising to heights of more than 150 feet. Its eastern terminus is Cape Cod, whose arcuate shape traces the boundary of an ancient lobe of ice. Stretching all the way to the state of Washington, the ridge passes just south of the

Great Lakes. The Great Lakes themselves interconnect to form what amounts to an inland freshwater sea that formed after the most recent glacial maximum. At first the lakes were ponded between retreating glacial lobes and the high ground to the south. They remain today in depressions the glaciers excavated. Similar bodies of water presumably formed after earlier glacial maxima.

Tundra with scattered spruce trees fringed the southern margin of the ice sheet. During the greatest expansion of the ice, this austere landscape trended across Pennsylvania through Ohio and Indiana to the Rocky Mountains in Montana. Even the southern United States was cooler; southern Florida, for example, ceased to be tropical. At the same time smaller glaciers expanded in mountainous regions. Their remnants still sit high in many valleys of the Rockies. The lower reaches of the valleys generally remain smooth and U-shaped. When the glaciers expanded and crept toward lower ground, they scoured the valley walls. Mountain glaciers grew even in the Southern Hemisphere, far from the great ice sheets of the north, and they persist today near the equator in valleys that scar Africa's tallest peaks. Early *Homo* roamed nearby during glacial maxima, when rivers of ice descended farther toward the Serengeti Plain.

Surface waters of the ocean that cooled when glaciers expanded yielded little moisture to the atmosphere, and on balance climates became more arid around the globe. There were exceptional regions, however, the most notable being the Great Basin of the American West. What happened there was that continental glaciers to the north of the Great Basin disrupted wind patterns, bringing moisture up from the south. Rainfall was plentiful in areas that are now desert, and lakes dotted the landscape. When Brigham Young descended from the Wasatch Range in 1847 and viewed the vast body of water below, he declared to his Mormon followers, "This is the place," or words to that effect. He actually was twenty thousand years too late, gazing at a lake far saltier than the ocean. At the height of the most recent glacial maximum, the predecessor of the Great Salt Lake had been fresh and had covered a third of the present state of Utah. The modern lake is a condensed remnant of this ancient Lake Bonneville, whose shorelines are now high and dry, incising slopes of the mountains that once confined its waters. At times the surface of Bonneville stood more than a thousand feet above the present level of the Great Salt Lake.

During glacial maxima continental glaciers rose to mountainous heights, in places reaching thicknesses of more than a mile. Mount Monadnock reveals precisely how high the glaciers stood in New Hampshire. They smoothed the lower slopes of the mountain as they ground past, but the peak remained rugged, an island standing in an ocean of ice. Modern visitors can easily detect the line of demarcation between the scoured and unscathed portions of Monadnock. Even before the idea of continental glaciation had gained general acceptance in the nineteenth century, scientists recognized its implications for the level of the seas. As glaciers locked up precipitation that would otherwise have found its way back to the ocean via streams and rivers, the oceans would have shrunk. Estimates of the volume of glacial ice now indicate that sea level should have dropped at least three hundred feet during the glacial maxima and that shorelines should have receded close to the edges of the continental shelves.

The Ice Age World

We still live within the Ice Age, but we are positioned in an interval between two glacial maxima. One occurred about 20,000 years ago, and the other will take place about 70,000 years hence. Some earlier glacial minima have produced even warmer conditions than those of the present. During the most recent of the previous glacial minima, about 120,000 years ago, glaciers melted back a bit beyond their present total volume and the seas rose to a position about thirty feet above their present level. Many years ago, for my undergraduate thesis, I studied the origins of the Key Largo Limestone, a fossil coral reef that forms the upper Florida Keys and that dates back to this last glacial minimum. Quarried slices of this magnificent rock, displaying brain corals and other elegant fossils, ornament facades of buildings in the Miami area. The uppermost portions of the ancient coral reef, which now stand about twenty feet above sea level, were bathed in seawater when the reef was alive.

In many parts of the world, climates were slightly warmer during this pronounced glacial maximum than they are at present. Fossils prove that species such as the water chestnut and the marshland beetle, which today are restricted to more southerly regions, then occupied the British Isles. An extinct variety of hippopotamus also somehow made its way to Britain and spread throughout England and Wales; paleontologists have exhumed its remains even in London, from river deposits beneath Trafalgar Square.

Glacial maxima have also varied in intensity. The most recent was probably somewhat more pronounced than the glacial expansions that occurred close to two and a half million years ago, at about the time *Homo* evolved. Even so, we have become so familiar with world geography during the most recent glacial maximum that it serves as a useful model for understanding the kinds of things that happened when glaciers first spread to huge dimensions at the start of the Ice Age.

The recency of the latest glacial maximum puts it within the range of radiocarbon dating. Innumerable organic materials, including seeds and fragments of wood on the land and debris from algae in the oceans, have yielded dates very close to twenty thousand years, which means that they date to the time when glaciers were at their fullest extent. In the 1970s a group of scientists launched a large project, named CLIMAP, to reconstruct the distribution of climates and life for this key moment of geologic time. CLIMAP studies of fossil plankton indicated that oceanic temperatures were slightly lower than they are today, and more recent studies have shown that in the tropics they may have been as much as 4°C (7°F) lower than today. Cooling was, of course, much more severe at high latitudes, close to the glaciers. Around most sectors of Antarctica, sea ice extended some ten degrees of latitude farther north than it does today.

For life on the land far away from ice sheets, the drying of climates was probably as important as cooling. More arid conditions resulted mainly from the strong temperature gradients that extended from the tropics to the frozen polar regions. Temperature gradients produce strong winds. Anywhere in the world where winds drag shallow waters away from a coastline, deep, cold water rises to take its place. During glacial maxima strong winds intensified such upwelling in many regions. Cooler waters near continents meant less evaporation, and air masses that moved landward were drier than before. The results are especially evident in Australia, where dunes expanded and migrated actively under the influence of parching winds. Many of the dunes that formed about twenty thousand years ago were later stabilized by vegetation as wind speeds diminished to present levels.

Climates and Vegetation of Modern Africa

What about Africa, the crucible of human evolution? Before examining the environmental history of the continent in order to shed light on the fate of *Australopithecus*, we should review its modern geography. Africa today displays a symmetry that, while remarkable, is also predictable from what is known generally about climates and life on the planet. Rain forests occupy most of the equatorial region, and two deserts, the Sahara in the north and the Kalahari in the south, occupy zones that stretch from about 15 degrees to 35 degrees from the equator. Savannas—grasslands with scattered trees—lie between the rain forests and the deserts, in zones where rainfall is moderate but seasonal. Near the rain forests, where precipitation is greater, savannas give way to a narrow zone of transitional woodlands, in which trees cluster more densely and grass grows only sparsely in their partial shadow.

The heavy precipitation that gives rain forests their name falls in the vicinity of the equator for a good reason. Here the sun's rays strike the earth most directly, and the heat that builds up warms and expands the air close to the ground. The warm, rarefied air rises and then cools as it moves away from the strongly heated earth. Cold air holds less water vapor than warm air, so that the rising air eventually loses much of its moisture. The rains that result sustain tropical rain forests, which in Africa lie mostly within about five degrees of the equator.

The heavy precipitation that sustains the African rain forest also indirectly creates the dry conditions of the savanna nearby. The equatorial air that rises, cools, and loses its moisture over continents such as Africa builds up high in the atmosphere to form a desiccated mass of air that spills to the north and south. Having cooled and become denser, as well as drier, it then sinks. As the air approaches the warm earth again, it heats up, so that its relative humidity drops even lower. Under the influence of the earth's rotation, the descending air twists to the west and becomes the fabled trade winds. The desiccated air of the trade winds parches the land, producing deserts in zones centered about twenty-five degrees both north and south of the equator. In Africa these are the Sahara and the Kalahari.

The savannas that are positioned between the rain forest and the deserts that lie to the north and south receive only seasonal moisture because the rainy equatorial zone shifts northward during one part of every year and southward during the other part. The Northern Hemisphere experiences summer when the earth's axis of rotation aims the North Pole toward the sun. At this time, between June and August, the rainy equatorial zone shifts to the north and the northern savannas turn green; the southern savannas cool down and suffer drought. The tables are turned in December, when the rainy equatorial zone migrates deeper into the Southern Hemisphere and brings rain to the southern savannas for about three months, while those in the north turn brown.

This pattern of rainfall and vegetation is entirely predictable from global climatology. A regional peculiarity, however, is the absence of rain forest from the lowlands of easternmost Africa: rain forests extend from the Atlantic only to the rift valleys, some seven hundred miles from the Indian Ocean, where they give way to grassy terrain. This terrain owes its presence to the seasonal drought in the eastern equatorial region — drought that results from strong regional winds which cause the rainy zone to shift so profoundly every year that no region is wet year-round. When it is winter in the Northern Hemisphere, powerful cool, dry winds sweep down from the parched Arabian peninsula, pushing the rainy zone entirely south of the equator. When it is summer in the north, the Himalayan plateau to the east enters the picture. This immense upland heats up in the summer sun, warming the air above it. The rising air pulls the famous monsoonal rains into Southeast Asia from the Indian Ocean. So powerful are the monsoonal winds that they drag the rainy zone of nearby East Africa to a position fully north of the equator. In other words, during one part of the year the rainy zone of East Africa lies to the north of the equator, and during another part it lies to the south. The combined influence of the Arabian peninsula and the Himalayan plateau leaves no lowland area of easternmost Africa with abundant moisture throughout the year. Here, there is no rain forest.

Where there *is* African rain forest, to the west, much moisture is recycled. It evaporates from the lush jungle and then showers that same jungle again as the air that carries the moisture aloft cools down. There is still a need for a rich supply of moisture to replace that which rivers and winds carry away from the system. The source of most of this moisture is the nearby Atlantic Ocean, from which monsoonal winds continually push inland to replace the rising equatorial air. These humid winds migrate somewhat—northward during the northern summer and southward during the southern summer—but they never abandon the equatorial belt altogether. They are what sustain the rain forest.

Ice Age Desiccation

One of the most profound environmental effects of the Ice Age has been a reduction of rainfall in many regions of the world. In fact, during glacial maxima broad areas of Africa have become drier than they are today. During these arid intervals dunes that at present are defunct and stabilized by vegetation have become active in a zone that extends about three hundred miles south of the present Sahara. As dunes are wont to do under the influence of desert winds, they have migrated. The same pattern is evident south of the equator, where ancient sands of a previously expanded Kalahari Desert now lie buried beneath grassy savannas. Although most of these ancient dune deposits remain to be dated, they obviously represent glacial maxima.

Climatic patterns of twenty thousand years ago serve as a general model for Africa during glacial maxima. During this most recent pulse of heavy glaciation, the levels of many large African lakes dropped far below their present positions, reflecting conditions that were more arid. The descent of glaciers down mountain valleys reveals that climates were cooler as well as drier. Early observations that montane glaciers exist in Africa today met with much skepticism. In 1948 a member of the Royal Geographical Society in London denounced as nonsense reports of such glaciers near the equator. One of the peaks that does, indeed, support glaciers is Mount Kilimanjaro, whose name means "shining mountain," apparently in reference to its glistening icy summit. Furthermore, bouldery moraines drape across valleys that incise Kilimanjaro at elevations fully three quarters of a mile below the limit of modern glaciers. Expanding glaciers pushed these deposits into place during the last glacial maximum. In the second half of the nineteenth century, these lowly tills convinced knowledgeable geologists that the chill of a glacial interval not far back in the history of the planet spread even to equatorial latitudes.

In the twentieth century geologists have expanded the story of glacial effects to the marine realm. CLIMAP data show that marine plankton underwent dramatic changes off the west coast of Africa during glacial maxima. Offshore from equatorial Africa, the modern Atlantic remains near 28°C (82°F) throughout the year. Microscopic fossils of plankton in sediments of the deep sea indicate that, twenty thousand years ago, during the most recent glacial maximum, temperatures there dropped to about 23°C (73°F) in February and rose only to about 25°C (77°F) in August. In other words, the average temperature was colder than today by about 4°C (7°F). This cooling resulted not so much from the cooling of air currents as from the strengthened upwelling of frigid waters.

Upwelling actually results from the horizontal movement of water. A great counterclockwise current cycles perennially round the South Atlantic under the influence of the earth's rotation, dragging coastal waters away from Africa. Cold waters well up from the depths to replace those that flow seaward. Upwelling is strongest in a zone between about twenty and thirty degrees south of the equator, but the gyre drags some of the cool waters that surface there northward, toward the equatorial zone, and then carries them westward across the Atlantic. The even cooler conditions off western Africa twenty thousand years ago reflect stronger upwelling. This is hardly surprising, in that cooling of the poles intensified temperature gradients between these regions and the equator, so that winds inevitably blew more strongly. Large masses of cool water then rose from the depths and flowed to the equatorial zone, yielding less moisture to the monsoonal winds than is provided by the warmer waters that border Africa today. This explains why large tracts of Africa became drier when continental glaciers expanded.

Stronger seasonal contrasts must also have come into play, leaving less of equatorial Africa rainy throughout the year. The strengthened trade winds must have pushed the warm, rainy zone farther north during the northern summer and farther south during the southern summer. This left only a narrow central belt moist year-round and capable of sustaining the rain forest.

Decimation of Rain Forests

Indeed, it appears that the drier conditions of glacial maxima have repeatedly shrunk the African rain forest quite drastically. Again and again, the watering system of this lush jungle has broken down. The modern rain forest stretches from the Atlantic to eastern Zaire, which is closer to the Indian Ocean. During glacial maxima, it seems, the rain forest shrank to three small patches, one near each end of its present extent and the third in between, in southern Nigeria and Cameroon. These three oases sometimes added up to no more than about 20 percent of the present extent of the African rain forest.

The present geographic distribution of plants and animals is what reveals the severity of the environmental crisis that the Ice Age has repeatedly visited upon the rain forest. Two patterns stand out. One is the total restriction of certain biological groups to the three areas of high diversity—or even to just two of them. It appears that many species have not returned to broad regions from which their populations disappeared during the last glacial maximum. The second pattern is the concentration of the biological diversity within the three small areas identified as refugia. Far more species occupy these areas than intervening regions.

Gorillas illustrate the first pattern. The single living gorilla species, *Gorilla gorilla*, occupies only the central and eastern

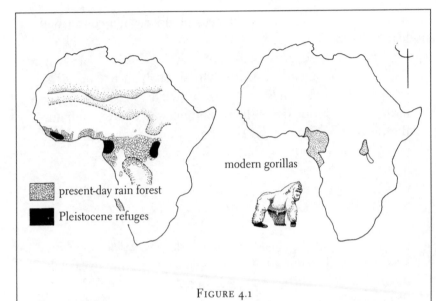

FIGURE 4.1

The relationship between the distribution of modern gorillas and changes in African environments during glacial maxima. The locations of the two subspecies of *Gorilla gorilla* correspond to two of the residual areas of rain forest that, other evidence shows, remained during glacial maxima when active sand dunes of both the Sahara and Kalahari deserts extended much closer to the equator (dashed lines) than they do today (solid lines).

African refugia. Furthermore, the two geographically separated populations represent distinct subspecies, or races. It would appear that the division of the gorilla populations has led to a small amount of evolutionary divergence. So far apart are the two populations that even during glacial minima, when the rain forest has grown up between them, they have not been able to migrate far enough to reestablish a single, continuous population.

Birds illustrate the second pattern. Each of the apparent refugia supports more avian species than any other area of modern African rain forest. In fact, may species of birds live within one or another of the refugia and nowhere else. The same is true for many groups of mammals, reptiles, amphibians, and butterflies. Plants also display this pattern: about eight thousand species are confined to a single sector of the present rain forest.

The implication that the three African regions are, indeed, isolated loci of survival gains strong support from the presence of similar geographic patterns in tropical rain forests of other continents. Both the Amazon rain forest and the jungles of the Far East also contain what seem to be refugial areas—pockets where species remain concentrated today, apparently having been confined there when rain forests shrank and fragmented in the recent geological past.

Climatic changes first shrank the African rain forests in the middle of the Pliocene epoch, at about the time the Ice Age began. The spread of windblown sand points to the expansion of arid conditions throughout most of northern Africa, and changes in fossil pollen shows that plants adapted to both cooler and drier climates extended their ranges in Ethiopia and Kenya. In fact, according to the analyses of the French paleobotanist Raymonde Bonnefille, the pollen reveals that climates in East Africa were even warmer and moister just before the Ice Age than they are at present and that when the Ice Age began they became both cooler and drier than today. Forests fragmented, while grasslands with only scattered trees and copses of trees expanded in the new, seasonally arid climate. Thus, the wrenching oscillations of the African ecosystems began, and the stage was set for the entrance of *Homo*.

CHAPTER FIVE

Death Comes for *Australopithecus*

We have glimpsed the before and the after—the Africa of *Australopithecus* and the Africa of the Ice Age. Why, once the forests had shrunk, did *Australopithecus* no longer fit the picture?

In contemplating the demise of *Australopithecus*, we must bear in mind that this creature, which flourished before the floral change, was by no means an arboreal acrobat comparable to a chimpanzee. It could not have swung through the high canopy of a rain forest. Instead, its populations would have depended on groves of smaller trees that grew under less rainy conditions at the margin of the tall forest. There, where numerous large predators also roamed, *Australopithecus* would have gathered food and found enough safety above the ground to survive.

The habitat favored by *Australopithecus* was probably what in Africa is sometimes termed the transitional woodland: a zone situated between the densely canopied rain forest and the savanna. Few trees of the transitional woodland are taller than fifty or sixty feet, whereas the canopy of the rain forest stands at two or three times that height. The crowns of the trees in the transitional woodland form only a thin, partial canopy, so that shrubs and sparse grass are able to spread beneath them from the neighboring savanna. The small trees of such woodlands would have been accessible to

Australopithecus. Within them our ancestor would have found fruits and pods to eat and forking branches on which to rest at night.

Today, the transitional woodlands of Africa are very narrow, averaging only about 150 feet in width, but this is a fairly new condition. Fire accounts for the abruptness of the transition between dense forests and savannas. Humans set range fires during the dry season to benefit agriculture within the savanna belt. These annual blazes destroy small trees and seedlings, turning what would otherwise be partly wooded grasslands into open, grassy plains dotted by occasional trees. Before populations of *Homo* began to set fires, about a million years ago, the transitional woodlands were almost certainly broader than they are today. Further, these likely habitats for *Australopithecus* would have been even more extensive before the Ice Age, when the annual rainfall declined less abruptly northward and southward from the equator than it does today. The gradient along which the rainfall declined would have been gentler because the deserts were not only smaller but also farther from the dense forests. Broad woodlands in between would have constituted an extensive, distinctive habitat rather than a narrow transitional zone like the one that today separates dense forests from savannas. It is easy to imagine that, until the Ice Age began, these vast woodlands offered *Australopithecus* ample refuge from predators.

As the Ice Age got under way, the woodland habitat of *Australopithecus* must have shrunk dramatically. In the first place, the woodlands must have narrowed as seasonally arid climates spread equatorward. In the second place, the total length of the woodlands, around the periphery of the rain forest, must have shortened as the rain forest contracted markedly. Carbon isotopes in soils support these conclusions. Thure Cerling of the University of Utah has shown that plants that conduct photosynthesis in different ways inject different proportions of carbon isotopes into the soils in which they grow. Soils beneath forests differ from soils beneath grasslands. In order to infer what floras were present where *Australopithecus* and early *Homo* lived, Cerling has measured isotope ratios for ancient soils at numerous sites that have yielded fossils of the human family. His results are spectacular. A major floral transition appears to have taken place about two and a half million years ago. Cerling's analysis indicates that all fossils of the human family older than this are associated with woodland or forest soils. Furthermore, only fossils younger than about two and a half million years ago are associated with soils of wooded grasslands. It is evident that human ancestors rather abruptly began to occupy open habitats, which, as the geological record of pollen and windblown sand indicates, were expanding markedly at about this time.

Carnivores and the Ice Age

Relationships between predators and their prey in modern Africa offer indications of the way in which the Ice Age should have created a crisis for *Australopithecus* by making it impossible for this relatively defenseless animal to rely on trees for refuge. Today, very few large African predators follow the migrating herds of hoofed animals as the dry season gets under way. Most adult predators are anatomically ill-suited for immense journeys, and their young also lack the wherewithal to travel great distances. Small groups of unattached male lions, known as nomads, do migrate with the herds, but they suffer high mortality before straggling back at the end of the dry season. The larger prides eke out a living during this trying time by moving into nearby woodlands, where they subsist on the sparse populations of buffalo, impala, and other herbivores that forage there throughout the year. Hyenas move their dens to the edge of the woodlands so that they too can hunt the herbivores that remain in nearby thickets and groves.

The dry season takes a terrible toll on carnivores. Owing to the

sparseness of herbivores in the woodlands and their even greater
scarcity on the open plains, many of the carnivores cannot ade-
quately feed their young. About two-thirds of all Serengeti lion
cubs fail to reach their second birthday. In fact, the great herbivore migration condemns many carnivores of all ages to starvation,

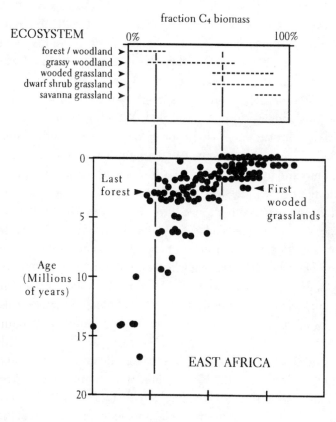

FIGURE 5.1

Carbon isotopes in ancient soils reveal types of ancient vegetation at sites where
fossils of the human family have been found. The last forests and the first wooded
grasslands at such sites date to about two and a half million years ago, when the
modern Ice Age began.

and, in doing so, it limits the sizes of carnivore populations. As for the migrating hoofed animals themselves, they not only escape from most predators by migrating during the dry season, but on their return they also meet fewer predators than they would have encountered if they had stayed among the predators to provide them with abundant food throughout the year.

We can employ what we know about the seasonal pattern of the Serengeti and about climatic changes in Africa during the past three million years to conduct what scientists call thought experiments. As a first experiment, let us consider where the predators should have lived before the start of the Ice Age and what impact they should have had upon herbivores. At that time forests and woodlands were more widespread than they are today, and open grasslands were more restricted. We know from the fossil record that most antelopes living prior to the Ice Age were adapted to wooded terrain; there were no vast migrating herds. Presumably, then, lions and hyenas, which today hunt extensively in woodlands during the dry season, thrived in these habitats year-round. This pre–Ice Age ecosystem should have been more stable than the modern one. Carnivores were not surfeited during a luxuriant wet season and then starved during a harsh dry season. The total population of herbivores must have been smaller than today but relatively stable. The perennial residents of the woodlands should have supported carnivores quite well; absent the bottleneck of the dry season, carnivores should have been more abundant than they are today.

A corollary of this conclusion is that, in the more stable ecosystem that existed before the Ice Age, predators should have cropped populations of herbivores so effectively that they controlled their numbers. Today, when herbivores evacuate the savanna in droves during the dry season, they leave many carnivores to starve. When there was no severe dry season and most herbivores remained in vast woodlands year-round, the populations of large carnivores should have expanded to the degree that they limited the numbers of hoofed herbivores. As the fossils from South African cave deposits suggest, *Australopithecus* also suffered substantial predation even before the cossetting forests shrank. In fact, throughout most of *Australopithecus*'s long stay on earth, predators may well have limited the sizes of its populations, just as they probably controlled the abundance of hoofed herbivores.

As a second thought experiment, consider what should have happened to African predators when forests and dense woodlands contracted at the start of the Ice Age. We know that many kinds of antelopes adapted to these habitats died out, but the fossil record shows that numerous antelope species adapted to open terrain then appeared. The question is, Should the array of African predators have been affected in much the same way? The answer turns out to be no. Of the five species of large predators in the modern Serengeti, only the leopard requires wooded terrain. This sparsely distributed, solitary cat needs only a small amount of cover for concealment, however. We can imagine that leopard populations declined as forests shrank, but not as severely as the populations of leaf-eating antelopes, which required vast expanses of trees and shrubs to supply them with leafy food. As the expanding dry season imposed a bottleneck on their food supply, lions and hyenas probably also declined in numbers to some degree, but these group hunters obviously took great advantage of the environmental changes as well, fattening up during the wet season on the vast herds that grazed on the luxuriant grasslands.

Given the general versatility of predators, then, we would not expect the start of the Ice Age to have devastated this group, because all that predators need are large herbivores to attack, whether these potential victims feed on leaves or grass. The fossil record confirms this conclusion. As far as we know, no species of large African predator died out as the Ice Age got under way. The three saber-toothed species disappeared *during* the Ice Age, but all were still alive to greet *Australopithecus* on the ground when the forests first shrank, about two and a half million years ago. The true sabertooths probably died out because they fed primarily on elephants, and elephants declined. At the start of the Ice Age, four elephant species, including a species of mammoth, occupied Africa, but only a single species, the one we call "the African elephant," survives today. As for the rest of the predators, the Ice Age did not bring extinction but only a life of ups and downs—a seasonal cycle of feast and famine.

The Demography of Extinction

Because a formidable array of large African carnivores lived on into the Ice Age, *Australopithecus* could not have endured the substantial increase in predation that must have occurred when the forests contracted. The problem lay in this large primate's pattern of reproduction. The pattern must have resembled ours and that of large apes, which are generally similar. A chimpanzee, for example, is born at a smaller size than a human baby, but it is more highly developed. Even so, chimpanzees reach sexual maturity more or less when we do. Chimps mature more rapidly in captivity than in the wild, where their diet is less rich. Jane Goodall found that males in nature tend to father their first offspring at the age of about fourteen and that an average female first gives birth at an age of thirteen or so. Gestation occupies about eight months, and single births are the norm. Twins, although perhaps slightly more common than in our species, are nonetheless rare. Female chimpanzees nurse their young for three or four years and average six years between births when they are rearing infants. In other words, a typical female produces very few offspring during her entire lifetime. So similar are chimps and humans in their low reproductive rates that *Australopithecus*, which shared traits of apes and humans, must have resembled these two living groups. It is unlikely that evolution took a detour toward some quite different pattern in the process of transforming an ape into a human.

A generation time of more than a decade is long for a mammal. Taking as long after birth as they do to begin producing at most one offspring a year puts chimpanzees and humans—and, presumably, *Australopithecus*—in a particular biological category. Animals of this kind tend to maintain stable populations by producing small numbers of offspring but then by raising them with such great care that, on the average, about one of the offspring survives into adulthood for every adult of the previous generation. Populations obviously shrink and expand, but usually less dramatically than do those of species that are at the other end of the spectrum of what ecologists call life history strategies. The species that contrast to large primates are known as opportunists.

Opportunists produce large broods at frequent intervals and suffer enormous infant and juvenile mortality. Such species are usually of small body size, which means that they grow rapidly to maturity. Rabbits exemplify this reproductive pattern to such a degree that they have attained metaphorical status. Heavy predation on opportunistic species is usually what takes so many young lives. When natural enemies are absent, populations of opportunists tend to explode, as when a plague of rabbits swept over Australia several decades ago following the introduction of the prolific animals to an island continent where they faced few predators.

For opportunists heavy mortality is a fact of life, offset by a high rate of reproduction. In contrast, less fecund animals, including large primates, survive by avoiding heavy predation. Events in Africa today show what happens to slowly reproducing animals when predation suddenly intensifies. Rampant human hunting now threatens the African elephant and black rhinoceros. Slow maturation and the birthing of solitary calves would doom these massive species to rapid extinction in the absence of legal protection, and it is not even clear that the preservation measures now in place will save them.

A similar shift of fortune must have befallen *Australopithecus* when the Ice Age began. A change in human behavior is what now threatens African pachyderms: an increase in the commercial value of the rhinoceros horn, which is falsely valued as an aphrodisiac, and of ivory, which has a unique decorative function. For *Australopithecus*, the problems of increased predation issued from a change in this animal's own behavior, albeit a change that fragmenting forests forced upon it. Of course, in Pliocene time there were no statutes to suppress the carnage that predators inflicted upon this human ancestor.

As copses of trees shrank and disappeared during early glacial expansions, the populations of *Australopithecus* faced an ecological crisis. As I observed in the previous chapter, even before the crisis they, like modern chimpanzees, must have exhausted local food supplies from time to time and moved from one area to another. A large grove of trees would have served as a home base until food ran out. We can imagine that they slept aloft and fed both in the trees and on the ground nearby, remaining close enough to a grove that they could find refuge there when predators threatened. It is not difficult to imagine the consequences for *Australopithecus* of the natural deforestation that occurred at the onset of the Ice Age. The expansion of grasslands between potential home bases must have forced populations to walk or trot farther than before across dangerous grassy terrain, where swift, powerful predators awaited them.

Can predators actually do in a species upon which they feed? Can they cause its extinction? The logical answer is that the predators can do so only if they have other species of prey to sustain them. Both theory and laboratory experiments indicate that if a single species of predator feeds on just one other species the two will become locked in an endless cycle of population expansion and decline. When the prey species is abundant, the predator will flourish, and it will crop the prey population to a low level. Lacking abundant food, the predator population will then crash, allowing the prey species to expand again. This will trigger another expansion of the predator species, which will then depress the prey species, and the cycle will continue, with the two populations oscillating slightly out of phase.

Seldom is the real world this simple, however. Most predators have a variety of prey species at their disposal, which means that the populations of both predator and prey need not expand and contract in concert. This must have been the situation in Africa at the start of the Ice Age, when shrinking groves of protective trees turned *Australopithecus* into an easy target for predators. Large African predators already had access to numerous species of prey. As they picked off members of *Australopithecus* and depleted populations of this large primate, their own populations should have remained large because they had many other sources of food, especially hoofed herbivores, to sustain them. In other words, they could have exterminated *Australopithecus* without having their own populations plummet.

It is important to understand that predators could have brought about the demise of *Australopithecus* without having hunted down the very last member of the genus—or even the last few individuals. The predators may simply have fragmented populations to the point where other factors finished the job. The outbreak of an infectious disease, a decline of food resources during a prolonged drought, a loss of vigor from inbreeding—events such as these might have caused the final extinction. When a species becomes exceedingly rare, even chance factors can snuff it out. Accidental deaths of infants or the failure of adults to find mates can make one sparsely distributed generation of a population the last.

Of course, *Australopithecus* somehow endured long enough in the presence of hyenas, lions, leopards, and the false sabertooth to give rise to *Homo*. The transitional population may have found a degree of protection in narrow forests that fringed rivers or in shallow caves beneath rocky overhangs. As far as is known, *Australopithecus* disappeared from the earth somewhat before 2.4 million years ago. The oldest known fossil remains of early *Homo* are just a bit younger, dating to about 2.4 million years before the present.

It remains possible, however, that one or more populations of *Australopithecus* survived for a short time after another population had evolved into *Homo*.

Long-lived Cousins

Robust australopithecines of the genus *Paranthropus* lived longer into the Ice Age than did *Australopithecus*, dying out something like a million years ago. These animals did not differ markedly from *Australopithecus* below the neck. What set them apart quite strikingly was their powerful jaws and broad, grinding molars. These features presumably expanded the diet of *Paranthropus* beyond that of *Australopithecus* to include such foodstuffs as tubers and roots. A broad diet may explain why the former genus fared better early in the Ice Age.

As copses of trees contracted and disappeared and dangerous open terrain expanded early in the Ice Age, *Paranthropus*, enjoying a wider range of food resources than *Australopithecus*, may not have needed to move so often from one protective copse of trees to another in a risky search for new sources of food. An ability to settle into one home base for a lengthy stay may account for the survival of *Paranthropus* further into the Ice Age. It may be no coincidence that the disappearance of *Paranthropus*, about a million years ago, roughly corresponded to the time when, according to Thure Cerling's isotopic analyses of soils, true savannas first appeared in Africa; before that time wooded grasslands were apparently the dominant form of vegetation to the north and south of woodlands and forests. Furthermore, isotopic records from marine foraminifera reveal that glacial maxima stepped up to a new level of severity at just about the same time. This was also the approximate time when *Homo* began to use fire, however, and perhaps the resulting increase in the frequency of range fires was even more important than the climatic shift in reducing the area of woodlands in Africa, to the detriment of *Paranthropus.*

Only Chimps and Homo Lived On

For chimpanzees, which of course have survived to the present, residual areas of rain forest presumably served as refuges during glacial maxima. These acrobats could find food not only on the ground but even high in the rain forest canopy, which was inaccessible to australopithecines.

Then there was the origin and flowering of the human genus. How did *Homo* actually come into being, and how did it manage to thrive in a world that had turned hostile to *Australopithecus?* The secret to success for *Homo* was obviously its prodigious brain. For this reason I will digress briefly from the narrative of human evolution to examine how our brain differs in form and function from the brain of an ape, and, presumably, from the apelike brain of *Australopithecus.* After this detour I will return to the Ice Age and the remarkably rapid emergence of big-brained *Homo* on the suddenly less verdant continent of Africa.

CHAPTER SIX

The Matter of Our Brain

If cerebral remodeling was the most important aspect of the origin of *Homo*, how did the brain of the new genus actually differ from that of *Australopithecus*? Having neither early *Homo* nor *Australopithecus* to examine in the flesh, we must turn to the closest living relatives of these two creatures for much of the evidence. *Australopithecus* was probably slightly ahead of a modern ape in brainpower, and early *Homo* was certainly behind modern humans. Nonetheless, we can learn much about the general trajectory of the change from the extinct to the living form by comparing the brains and brain functions of modern humans and apes.

The Human Brain in a Nutshell

The machinery inside the human brain is enormously complex. Complexity, however, does not increase "upward" uniformly from animals that we view as primitive to ones that we view as advanced. Here and there, as the tree of life has grown during the course of geologic time, intricacy has erupted along the way. The most complicated of all skeletal structures in the animal world is apparently the chewing apparatus of the lowly sea urchin, a spiny relative of the starfish. Known as Aristotle's lantern, this gadget consists of twenty skeletal parts and many muscles that, through their coordinated effort, enable the urchin to rasp velvety growths of algae from rocks and coral reefs. The human brain is, however, a far more complex organ from the standpoint of physiology, which encompasses not only mechanical movements but also biochemical processes. In fact, our brain is the most complex physiological structure on earth, and its existence is our sole justification for claiming that we stand at the top of the animal world.

Superficially, the human brain has the appearance of a great gray walnut. A narrow central fissure divides it, and its wrinkly surface manifests the intense folding, or convolution, of the cerebral cortex, which is the brain's soft outer shell. We cannot count, but only estimate, how many message-carrying cells, or neurons, occupy the cerebral cortex, but the number is something like ten billion.

The message of a neuron is electrical. A neuron receives a message by way of one or more dendrites, tentaclelike fibers that radiate from one of its ends. At the neuron's opposite pole is a fiber called an axon, which transmits the message to the dendrite of another neuron or of a muscle cell that then contracts.

Neurons expend a vast amount of energy to carry out their work. So large is the human brain that it consumes about 80 percent as much oxygen as all other parts of the body combined. To feed their large brain, humans must consume considerably more food than they would require if they were equipped with the brain of an ape. Neurons are, in effect, elite cells that are supported by minions known as glial cells. Glial cells supply glucose and other materials that the neurons require for their highly energetic activity. Beneath the gray matter of the cerebral cortex is a core of white matter, which consists mostly of supporting glial cells.

The Meaning of Brain Size

Whereas the brain of *Australopithecus* was the size of an orange, the modern human brain measures up to a good-sized grapefruit. The average adult cranial capacity for a chimpanzee is about a third as large as that for a modern human, but it is roughly seven-eighths as large as that for *Australopithecus*. What is the significance of the difference in volume of gray matter between chimpanzees and *Australopithecus*, on the one hand, and humans on the other? Inasmuch as the modern human brain is roughly three times as large as the brain of a chimpanzee, we might expect it to contain about three times the number of neurons, but this is not the case; the human brain has only about twice as many neurons. As Terrence Deacon of Boston University has observed, basic arithmetic explains this pattern. For a region of the brain to function as a unit, its neurons must interconnect. When there are many neurons in a functional unit, the addition of just a few more requires the formation of many fibers to connect them to other neurons, and these fibers must occupy space.

Because of its great size, the human brain comprises a vast network of neuronal connections. Variation in the number of key connections probably explains most differences in basic intelligence. It likely explains, for example, why some people with relatively small heads (and brains) are quite brilliant. Although such people may have fewer cortical neurons than people with larger brains, connection between neurons is what matters, and some people with relatively few neurons have many neuronal connections where they count the most. On the average, however, big brains contain both more neurons and more connections between them than do small brains. A species that has a much larger brain for its body size than another will be a quantum leap ahead in brainpower. Thus, because our brain is close to three times as large, humans stand far above both chimpanzees and australopithecines in basic intelligence, while we are only slightly ahead in body weight.

Body weight is important when considered in relation to brain size because much of the brain of any mammal serves simply to regulate bodily functions. Thus, on average, the bigger the body, the larger the brain must be. Among mammals there is a general statistical relationship between brain weight and body weight. In this relationship the brain does not keep pace with body weight during growth. In other words, a species that weighs twice as much as another, on average, has a brain that is less than twice as big. Brain weight lags behind body weight in accordance with a simple mathematical formula that, although easily computed, has yet to be fully explained. Humans stand apart from this relationship: for our body size we are eggheads.

Within a species, however, there is little relationship between brain size and body size. In other words, a large member of *Homo sapiens* is not especially likely to have a large brain, and a small person does not necessarily have a small brain. This is not a strange pattern; it holds for certain other organs as well, such as noses and penises.

The Vocal Limitations of Apes

What exactly can humans do that apes, with their much smaller brains, cannot? The answer is important to an understanding of the origin of *Homo* because, as I will discuss shortly, the brain of *Australopithecus* may not have differed markedly from that of an ape. Part of what we humans can do that is beyond the ability of all apes is to communicate with a language that employs logical grammar to connect verbal symbols in meaningful ways. Since Descartes argued in the seventeenth century that advanced language sets humans apart from other animals, many students of the mind have viewed the generation of complex language as the most important unique feature of the human brain.

Apes are physically capable of uttering very few of the sounds, or phonemes, that we humans combine into words. It was upright posture that facilitated the evolution of vocalization in our ancestors. With the head standing atop the torso, the larynx (voice box) was able to migrate down into an upright, stalklike neck in the course of our evolution to become our Adam's apple. In this position it is separated from the trachea (windpipe), and the resulting expansion of the throat provides ample room for linguistic movements of the tongue. A chimpanzee's tongue, in contrast, has little room to maneuver. The human throat structure is especially important for vocalizing vowels, as when we say "ah" for a doctor by depressing the rear of our tongue while emitting air through the pharynx.

The restructuring of the human throat for speech has not been without sacrifice. We have lost the ability to swallow and breathe at the same time, which can create problems. The epiglottis, a cartilage-supported flap at the base of the tongue, serves as a valve to keep food from entering the trachea when we swallow. Nonetheless, mishaps can occur, as when a bit of liquid from the mouth trickles down the larynx ("the wrong pipe") and causes distress or, much worse, when a chunk of food lodges there, necessitating the lifesaving Heimlich maneuver.

In apes the higher position of the larynx allows the epiglottis, at its upper end, to separate the windpipe from the esophagus during the act of swallowing, so that apes can use both tubes at once. Up to the age of about three months, human babies, like apes, can swallow liquids and breathe at the same time, because their larynx and epiglottis have not descended into the adult position. Humans can easily afford this aspect of development, which prevents the suffocation of suckling infants, because at such an early age the larynx need not be in a position for speech. Because speech is such a valuable adaptation, however, natural selection has produced the adult human configuration, despite the awkwardness and danger that it poses for breathing while eating. Here we see an obvious compromise in the origin of the human condition.

When this condition arose remains a mystery. We have no way of reconstructing the flesh and cartilage of *Australopithecus*, or any other extinct member of the human family, so as to learn to what degree it could have uttered the sounds of modern human language.

Simple Sign Language for Subhumans

As it turns out, chimps can actually learn some sign language—as long as humans provide them with rigorous training. In fact, they have learned simple systems of communication using not only American Sign Language for the deaf but also graphic symbols generated by computers and plastic tokens. Chimps have also learned to recognize spoken words. Sue Savage-Rumbaugh of the Yerkes Regional Primate Center in Atlanta, Georgia, taught the six-year-old Kanzi, a member of the rare pigmy chimpanzee species, to recognize more than a hundred words of English by sound.

Chimpanzees, like children, are better at understanding communications than at issuing them. Moreover, they learn the meaning of symbols only with difficulty, and their total capacity for such education is modest by human standards. In the first effort ever made to teach sign language to a chimp, Allen and Beatrice Gardner were able to endow an animal named Washoe with a vocabulary of only eighty-five signs by means of rigorous training over three years. In contrast, a child three or four years old, despite having spent much of its life in a prelingual state, can use as many as fifteen hundred words in speech and can understand two or three times this number. Furthermore, children simply assimilate vocabulary; they learn without the special training that chimps require in order to understand and manipulate symbols.

Chimpanzees are also unable to master complex syntax. They can learn the meaning of negative modifiers and of prepositions such as "in" or "on," but when it comes to phrasing, they can do little more than string two or three words together randomly, without mastering the systematic ordering that humans use to specify particular meanings. Chimps seem unable to compose questions, for example. Most of their constructions simply connect a verb to an object or to a subject and an object, and most of these crude communications are directed at some form of self-gratification. It turns out that dolphins, which are members of the whale family, can generally match chimps in the ability to understand simple sentences that are conveyed by signs or auditory signals. Given the size of its brain, *Australopithecus* probably had little more aptitude even for sign language than does a chimpanzee.

The Creative Deficiencies of Apes

A variety of mental traits are unique to humans, and not all of them are tied to language. What are the general cognitive skills that separate humans from apes? One of them is self-awareness—an understanding of our individuality and our ability to entertain unique thoughts. A chimpanzee can learn to recognize itself in a mirror—as can an orangutan, though less easily. A gorilla cannot accomplish this feat at all, however, a simple fact which illustrates that apes, in general, have only a marginal capacity for self-recognition.

Michael Corballis of the University of Auckland sees "generativity" as the key trait separating humans from apes. This is the ability to assemble a finite number of basic elements, such as physical raw materials or words, into an infinite variety of manufactured products or statements. Corballis's concept of generativity includes most of what we encompass with the word "creativity."

Merlin Donald of Queen's University in Kingston, Ontario, Canada, has suggested that different types of generativity may have had different evolutionary origins. He views what he terms "mimetic" skills as being more fundamental to the evolution of the human genus. These are abilities to mimic or depict something by means of primitive vocalizations, facial expressions, postures, or movements. Such representations entail inventiveness. In our deep history they led to narrative language, in which spoken symbols replaced mime in the portrayal of past events.

Nonhuman animals, including apes, do not relate particular past events to one another in any coherent way. They live from one experience to another without constructing an oral history: they fail to accrete collective knowledge. Apes can pass along only very elemental information from generation to generation. Not only are they unable to describe particular incidents of the past, but they also can generalize only weakly from specific events. In sharp contrast to the history of apes, proliferation of knowledge has been the essence of human cultural evolution.

Primitive forms of signing presumably preceded spoken language in human evolution. I suggest that here, and in our advanced cognition in general, categorization is at the root of our unique abilities. We assign things to categories that become the units we label, define, and manipulate. We label them with gestures, words, or numbers. In the process of categorizing them, we diagnose their properties. We also order them in time; relate them to one another in cause-and-effect relationships; add, subtract, multiply, and divide them; and change their orientations and locations, at least in our minds.

Language greatly facilitates categorization by providing convenient labels for the things we see, hear, or feel. It also permits us to manipulate objects and categories readily in our mind's eye, which is to say, in our imagination. We often articulate ideas inside our heads with unspoken words.

Our generative activities are hierarchical. With advance planning, we can fabricate a variety of items and then assemble these intermediate units into larger, more complex structures. Thus, modules that a factory produces become the prefabricated buildings that architects design in advance. In our writings the ideas of individual sentences become the broader messages of entire para-graphs—messages that the author had in mind before putting pen to paper or before turning on a word processor. The chain of para-graphs logically forms a larger whole, the letter or article or book that the author had originally conceived, at least in general out-line. No other living animal can engage in such complex planning for hierarchical construction.

Human language is doubly creative. First, we create the sys-tem—the words and the syntax that relates these words logically. We then employ this system in limitless inventions. Language facilitates thinking. When I am concocting a new scientific idea, I tend to mumble each piece of the argument to myself to develop it fully and test the logic. By categorizing items and concepts and by re-lating the categories to each other, precise words and syntax add an-alytical rigor or expose flaws in reasoning. A skeletal idea sometimes disintegrates when I formulate it as a sequence of clauses.

Of course, humans develop some ideas with mathematics, but mathematics is actually a shorthand for language; any number or symbol can be turned into a word, and we can then state any equa-tion as a sentence. Mathematics requires special nonlinguistic thinking skills, however. The same can be said for graphical or artistic portrayals, which sometimes generate conclusions in the absence of the written word. Experiments also establish some of our knowledge, but these entail planning and a general under-standing of chains of cause-and-effect relationships that greatly ex-ceed the capacities of chimpanzees.

Apes not only learn slowly and show little capacity for storing vocabulary, but they also lack the ability to generate their own lan-guage. Special training has proven that communication by sign-ing is well within their range of motor skills, even if speech is not. Nonetheless, they are evidently too limited in brain power to in-vent even a simple form of sign language for themselves that ad-vances beyond isolated gestures.

Chimpanzees also display a relatively complex social structure, one in which cooperation between individuals is extensive. They are, for example, the only animals that engage in internecine warfare that goes beyond attacking territorial interlopers in the manner of lions and hyenas. For one thing chimps—mostly males—actively patrol territorial boundaries in order to spot potential invaders. In addition, they make group forays into enemy territory. Jane Goodall concluded that these raids actually seem designed to murder members of neighboring troops. Often, one member of the invading troop holds down a victim while others bite and pummel it, inflicting fatal wounds. The aggressors commonly drag the victim back and forth in an ostentatious manner that they seem never to employ when killing members of other species for food. Goodall actually observed a large troop's systematic extermination of a smaller troop by means of a series of gang attacks on the victims' home turf.

In such activities chimpanzees engage in much more advanced behavior than nearly all other nonhuman species of animals. Even so, the group behavior of chimps, aggressive or otherwise, is simple and stereotyped, and it is obvious that chimps have not been advancing appreciably during their history in the use of tools or in the development of cooperative social structures. By human standards they have stagnated at a very primitive level.

Australopithecus had a brain that was only slightly larger for the animal's size than the brain of a chimpanzee. The fact that this animal produced no stone tools or other artifacts durable enough to have left a fossil record confirms that its mental capacity was more apelike than human. Like chimps, *Australopithecus* probably passed information from generation to generation by example rather than by symbolic description entailing words or signs. The offspring of *Australopithecus* probably learned from parents simply by imitating their behavior.

How the Complex Human Brain Takes Shape

How does a human grow the brain that is so much more powerful than that of a chimpanzee? Most neurons that remain in the adult brain form between three and five months after conception, while humans are small embryos. Neurons proliferate more slowly for some time after this early stage of rampant multiplication, but virtually none are added after birth. In fact, our embryonic growth oversupplies the brain with neurons. Many are weeded out as the brain grows during infancy and childhood. The key additions to the brain after birth are not neurons themselves but the fibers that connect them. Although some of these axons and dendrites make only temporary connections and then atrophy, there is a great overall increase in the number of connecting fibers early in development. The result is a complex network of neurons. Neurons also migrate to new locations as the brain matures, and glial cells, which nourish the neurons, continue to proliferate. The degeneration of neurons is, in fact, a normal feature of human development. Following a kind of "use-it-or-lose-it" principle, neurons and fibers that fail to connect with functioning target cells die out. In some regions of the human brain, most of the neurons present at birth expire during the first few months thereafter.

The death of neurons and the differential growth and death of neuronal fibers produce discrete networks of neuronal connections that Gerald Edelman of Rockefeller University refers to as maps. A map performs a function. The connections between its neurons have survived because they have been put to use. A map may record a memory, or it may encode a particular pattern of thought or bodily movement. Genetic controls constrain the broad pattern of brain growth, but the uses to which neurons are put influence the pattern by which connections develop. Further, maps are not fixed but dynamic. Thus, we add facts to the body of knowledge stored in our heads. We also forget details, because memory decays in the course of time as connections between neurons change. Marian Diamond of the University of California at Berkeley found that environmental stimulation plays a powerful role in the proliferation of neuronal connections. She and her students allowed young rats to intermingle in a cage that challenged them to navigate a maze in order to find food. The animals grew brains with as many as 50 percent more neuronal connections than did young rats that were isolated in darkened, soundproofed cages. Clearly, the growing brain needs exercise in order to approach its full potential.

The Structure of the Adult Brain

In order to understand how the human brain differs structurally from that of an ape or australopithecine, let us examine its various parts. Edelman characterizes the human brain in a somewhat nontraditional way, but one that makes a great deal of sense. He divides it into two functional parts. The more primitive part includes the brain stem and limbic system, which constitute the core of the brain. The brain stem, at the very base, connects with the spinal cord, and the limbic system consists of several components that overlie it. These various nuclear structures govern many patterns of behavior and physiology that are beyond conscious control: heart rate, sleeping, sweating, unconscious breathing, appetite, digestive functioning, sexual desire, and impulsive reactions to danger. The more advanced, outer region of the human brain includes the thalamus, positioned between the brain stem and limbic system, as well as the cerebral cortex, which is the wrinkled shell of the brain. This outer region of the brain receives signals from the environment—by way of sight, hearing, smelling, taste, and touch—and generates responses. It also categorizes and manipulates the items and events that we sense, re-create, or imagine: it is the seat of our intellect. This so-called cortical system evolved later than the brain stem and limbic system. The growth of our brain crudely mimics this evolutionary pattern. The brain grows upward and forward from the brain stem. The cerebral cortex is the last major part to form, mantling the more primitive regions and endowing the brain with its prodigious overall size. So important is the cerebral cortex in advanced animals that it is known as the neocortex, or "new cortex."

Particular regions of the neocortex have particular roles, although each of these regions operates in concert with other portions of the brain. It is important to understand how these regions function in order to understand how the human brain evolved. The neocortex is divided into a left and right hemisphere, and each hemisphere is divided into four major lobes. The large frontal lobe extends forward from the central region, to underlie the forehead and meet the rear sector of the eyeballs. Behind it sits the parietal lobe, and still farther back and forming the rear of the brain is the small occipital lobe, which houses visual functions. The temporal lobe, extending along the side of the brain in the region of the temple, is in contact with the other three lobes. Distinctive zones on a smaller scale are the regions of cortex that receive sensory information or that stimulate movement of specific body parts. The sensory cortex, which receives signals, is a narrow zone of tissue positioned like a headband along the front margin of the parietal lobe. Just in front of the sensory cortex is the motor cortex, which borders the frontal lobe, like a second headband. The location of these zones is no accident. For quick response to signals, the motor region for a given body part lies adjacent to the corresponding sensory region.

The sensory cortex and motor cortex of the left hemisphere serve organs on the right side of the body, and comparable zones of the cortex of the right hemisphere serve body parts of the left side. The site for the toes lies at the top of the brain, along the base of the crevice that separates the left and right hemispheres. Next below this comes the region for the ankles, followed in sequence by the regions for the knee, hip, trunk, shoulder, elbow, and wrist.

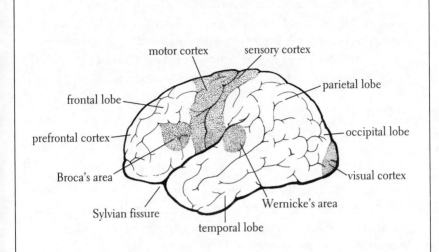

FIGURE 6.1

Major areas of the left hemisphere of a typical human brain. In some people's brains Broca's and Wernicke's areas, which function in the use of language, are positioned in the right, rather than the left, hemisphere. Thinking is centered in the forward region of the frontal lobe.

Partway down the side of the brain are regions for the hand and each of the five fingers, ending with the thumb. The region of vocalization—which serves the lips, jaw, tongue, and throat—sits far down along the side of the neocortex, close to the temporal lobe. This location has special significance. It places the control system for speaking close to two other regions of the brain that play central roles in the use of language.

At the front of the brain is the prefrontal cortex, sometimes termed the frontal association cortex, because it is the place where we associate various kinds of information logically in order to make decisions: quite simply, this is where we think. In the evolution of the human skull from that of *Australopithecus*, one of the most striking changes was the transformation of a frontal region that was low and sloping, like that of an ape, into our nearly vertical forehead. This change reflected a great expansion of the prefrontal cortex, which must have given rise to the thinking power that makes us so much smarter than an ape or australopithecine.

Two Brains in One

As it turns out, the left and right sides of the human brain each perform certain unique functions. A surge of interest in this so-called left-brain–right-brain dichotomy has led to the publication of numerous scientific articles and books in recent years. What was the origin of this lateralization? Why did it evolve?

Because verbal language is of paramount importance to the human species and is also unique to us among all living animals, much research has focused on the way in which our brain concocts and deciphers speech. Fascination with speech has, in turn, heightened scientific interest in the fundamental asymmetry of the human brain, because for most people—between 90 and 95 percent—the left hemisphere plays the dominant role in encoding and decoding language. The two key areas lie near the left Sylvian fissure, a cleft that separates the temporal lobe along the side of the head from the sector of the brain that lies above it. The anterior region that is devoted to language bears the name of the French physician Paul Broca, who in 1864 reported that patients who had suffered damage to the back portion of a particular convolution of the left frontal lobe, just above the Sylvian fissure, experienced loss of speech. Soon thereafter the German neurologist Carl Wernicke discovered that damage to a region of the left temporal lobe, to the rear of Broca's area, created difficulties in speech comprehension. It turned out that a bundle of fibers connected Wernicke's area, where we seemed to receive and make sense out of speech, and Broca's area, where we seemed to produce it.

Wernicke's area, which is positioned slightly above the ears, lies not far from the sensory region of the brain that processes sound, and Broca's area is close by the motor area that controls the mouth. Thus, it once seemed that a sector of the left brain was neatly partitioned into an intake and an output system for speech. The situation turns out to be more complicated, however. Recent observations have implicated the rear portion of Broca's area in the understanding of speech and the front part of Wernicke's area in its production. In other words, the two areas share a pattern of operation: they seem at least partly to duplicate one another in governing language skills. Furthermore, positron emission tomography (PET) scans and nuclear magnetic resonance imaging (NMRI), which measure blood flow and, hence, activity in the brain, reveal that many areas of the cortex and even of deeper portions of the brain play a role in the production and comprehension of speech. The left cortical region is of primary importance, but it does not do the job alone.

Furthermore, when we interpret emotional intonations that add nuances to speech, the right hemisphere comes heavily into play. In fact, this half of the brain appears to deal with emotion to a greater extent than does the left hemisphere. It is interesting that most of us express feelings more strongly on the left side of the face, which the right hemisphere controls, than on the right side. The right hemisphere also dominates in the perception of spatial relations and in holistic thought in general. Its activities must coordinate with the left hemisphere when language entails an organizational perspective of one sort or another. On the other hand, the left hemisphere usually contributes more than the right to sequential arrangements, as when we generate strings of words and phrases in speaking or when we employ a particular series of bodily movements to accomplish a task.

The human brain is asymmetrical not only in its functioning but also in its physical form. A comparison of the two hemispheres reveals that the front of the right one projects farther forward in most individuals, whereas the rear of the left hemisphere projects farther backward. In addition, the Sylvian fissure is usually larger in the left hemisphere, and the left planum temporal, which connects with Wernicke's area, is also larger. The location of the latter two asymmetries suggests that they are related to language functions in which the left hemisphere dominates. It is interesting that in most dyslexic humans the planum temporal is roughly as large in the right hemisphere as in the left.

Many hypotheses have been advanced to explain the origin of our brain's asymmetry. There has been a widespread view that the development of advanced communication was the most important facet of our ancestors' cerebral ascendancy above the rest of the animal world. This has led to much theorizing to the effect that our cerebral asymmetry evolved in association with the advent of speech or perhaps with the origins of sign language which our ancestors used before they engaged in syntactic vocalization. The problem with this line of thinking is that recent research has uncovered cerebral asymmetries in lower mammals that in many ways resemble those of humans.

The brains of apes are lopsided in the same way as ours, with the right hemisphere projecting forward and the left hemisphere, to the rear. Also in chimpanzees, as in humans, the Sylvian fissure is longer on the left side of the brain than on the right. Asymmetry in these apes is not only a matter of physical form. Chimps trained to recognize symbols respond more quickly to ones revealed to their right eye, which the left side of the brain serves, than to symbols revealed to the left eye.

Cerebral asymmetry actually extends to much simpler animals than apes. Experiments have shown that monkeys whose left hemispheres are damaged have difficulty distinguishing between various types of vocalizations made by other monkeys. No such effect results from damage to the right hemisphere. Female mice also record the ultrasonic peeps of their offspring primarily in the left side of their brain, and several kinds of birds generate songs primarily in the left brain.

For animals in general, it appears that the left hemisphere tends to dominate in producing and processing vocalizations. Even so, we cannot be certain that a single evolutionary thread connects all species in this regard. The pattern may have evolved separately in different groups of animals, and it may not hold for all kinds of animals. Also, the pattern is not rigidly fixed within species. For about one in ten right-handed people and a third of left-handers, the right hemisphere, not the left, is dominant for language.

Many theorists have tried to establish a fundamental link between the asymmetry of the brain and handedness. Dexterity, of course, derives from the Latin word for "right," because most of us are right-handed when it comes to manual precision. It has seemed significant that the left part of our brain serves our right hand as well as governing our use of speech. The two skills—manual and linguistic—are combined in the use of sign language. As would be expected, damage to the left side of the brain turns out to impair the signing abilities of people previously proficient in American Sign Language. Lesions to the right side of the brain have a complementary effect, diminishing victims' abilities to interpret the spatial significance of signs that are directed at them. These observations have spawned suggestions that it was not the origins of spoken language that led to the lateral partitioning of the human brain but the earlier advent of sign language—communication in which the right hand took the lead and the left hemisphere, which controlled this hand, took on the role of organizing its complex, sequential gestures. As I will discuss shortly, it may be that *Australopithecus*, with a brain slightly larger than an ape's, used simple sequential gestures.

We might still ask why one hand should have specialized in such gesturing. Why should both hands not have been used with equal facility, to afford greater versatility or to generate more sophisticated "syntax"? More generally, why should our two hands not have become equally adept at precise movements and manipulations?

A group of scientists led by Peter MacNeilage of the University of Texas has offered an ingenious hypothesis to explain how right-handed gesturing might have arisen in our primate ancestors. These scientists and others have found that monkeys and other primitive primates tend to use the left hand for reaching and grasping. If this specialization evolved as a preference for postural support—for clinging to a branch or standing on three legs—the right hand would have remained available for handling food and gesturing. In this manner an early primate might have developed differing specializations for its left and right limbs. The issue of versatility does not disappear entirely in this scenario, however. Dexterity in both hands has advantages.

In fact, handedness has never been satisfactorily explained. Clearly, ambidexterity would enhance our performance in many tasks. One view is that our handedness endows us with a sense of direction—a means of knowing left from right. People who happen to be ambidextrous are not seriously disoriented, however—at least not to such a degree that the resulting problems would necessarily offset the potential benefits of two-handed dexterity in a primitive society.

Humans use a fair portion of the brain in conducting complex manual tasks and verbal communications. Might it be that when evolution introduces entirely new capacities to the brain it tends to establish them on one side in order to conserve preexisting, and still essential, capacities of cortical tissue on the other side? I would venture this as an explanation for brain lateralization that seems to have eluded neurophysiologists. This explanation has the advantage of applying even to simple creatures. In other words, it is not linked specifically to the asymmetry of the human brain, which may, after all, have grown out of less pronounced asymmetries in lower animals.

My proposal is best understood by picturing a perfectly symmetrical ancestral brain and then considering what must happen when the capacity for some new kind of function evolves within it. A fundamental problem is that the new function must displace a preexisting function: a region of the brain that played a particular role must convert to a new one. Simply put, if this change takes place in only one hemisphere, the original function can be totally conserved in the other hemisphere. For example, it seems likely that the regions of the human left hemisphere now devoted to language had previously joined comparable regions of the right hemisphere in the management of spatial relations, holistic concepts, and emotions. The spatial, holistic, and emotional functions have been conserved on the right side of our brain.

If the brain grows larger while a new function supplants an original function on only one side, then the original function need not suffer. If, for example, the brain doubles in size, then the amount of tissue devoted to the function on the side that conserves it will equal the total amount of similar tissue that existed when the brain was half as large and the function was distributed equally between the two sides. Thus, the great expansion of the human brain beyond the level of *Australopithecus* may have conserved key functions effectively on one side of the brain while altogether new functions related to things such as speech displaced them on the other.

To put it another way, asymmetric positioning made it possible for brain expansion not simply to amplify existing functions but also to make room for altogether new ones. The evolution of new functions would otherwise have been blocked because they would have displaced essential existing brain functions.

Previous workers have suggested an explanation for cerebral asymmetry that sounds similar to the one I have outlined but is, in fact, quite different. They have claimed that evolution has saved brain tissue by positioning certain functions primarily on one side instead of wastefully duplicating them on the other side. The logical flaw here is that natural selection would not have produced more tissue than necessary simply because the tissue was divided into two portions. Natural selection is parsimonious. It would simply have established on each side half of the total mass of tissue required to perform the function.

A Single Function Engages Many Regions

There may be more advantages to lateralization of the brain than the conservational one I have suggested. Lateralization may, for example, speed up certain cerebral operations by positioning two or more regions of the brain that interact in these operations close to one another so that the travel time for messages along nerve fibers is automatically shortened. Even so, new imaging techniques reveal that many distant areas of the brain coordinate efficiently during particular brain activities.

Innovative scanning techniques that highlight blood flow in the brain are now illuminating this astounding organ for the first time, just as the telescope brought vast tracts of the universe into view in the seventeenth century. Objects that occupy different parts of the visual field, for example, turn out to fire off neurons in different regions of the brain. In effect, we have in the visual cortex at the back of our brain a coded map of what we have seen. Quite different regions altogether deal with movements that we observe, and when color is part of our view, particular regions of cortex farther forward come into play. It is perhaps no surprise that when we dredge up an image from memory the areas of the brain that recorded the image in the first place take part once again; obviously, the image was mapped onto them. Wernicke's area may play a central role in the interpretation of language we read, but particular sites farther back in the left hemisphere become active when the eyes view actual words rather than meaningless arrays of consonants; the visual system selects particular material from what we see—material that can be understood as language.

Even if the left hemisphere dominates in language functions, the right hemisphere interprets intonations that give subtle meaning to speech. The right hemisphere also relates language to spatial and conceptual frameworks. PET scans have revealed that even the lowly cerebellum plays a role in the use of language. The cerebellum is a small, globular structure that sits behind the brain stem at the base of the brain. One of its major functions is in programming instructions for complex sequences of movement that we carry out frequently as a matter of habit. It is perhaps no great stretch to imagine that natural selection put this particular capacity for programming to use for encoding patterns of vocalization which are frequently repeated.

In general, the new PET scan and NMRI techniques are showing that the brain is not neatly divided into discrete regions that function independently. Particular regions play particular roles, but they act in concert with others. In fact, we perform some tasks most efficiently when both hemispheres of the brain take part. Efficiency is especially problematical when we confront moderately complex tasks. For example, when we must recognize patterns with the right eye alone and then respond with the right hand via a signaling device, we are using the left side of the brain for both functions, and our reaction time turns out to be relatively slow. When similar patterns are instead presented to the left eye, the right hand responds more quickly, apparently because the overall task is divided between the two sides of the brain. Thus, the two-step sequence of processing the information and then activating a response entails a slight delay when it is confined to one side. How general this kind of advantage may be is unclear, but its mere presence suggests that the packaging of the intake and output components in close proximity, in the left hemisphere, may not function to speed up our linguistic activities.

Trying to Read Skulls

In past centuries phrenology enjoyed credence. Practitioners of this pseudoscience claimed the ability to read the qualities of people's minds by measuring the shapes of their skulls. Legitimate experts actually can infer from patterns on the inner surfaces of skulls some of the larger features of primate brains—features that reflect differences between the brains of different species, for example. Even here, however, the possibilities are limited because the configuration of a brain's surface is not well imprinted on the interior of the skull. Because of this limitation, fossils offer meager information about the brain of *Australopithecus*. Its brain was lopsided front and back like ours, but we cannot infer much more from the faint impressions on the interior of its skull. The convex convolutions of the brain and the depressed grooves, or sulci, that separate them are generally unrecorded on fossil skulls. Even the Sylvian fissure, the deep groove in the side of the brain, fails to show up in the overlying bone. Patterns of the brain's surface are better imprinted on juvenile than on adult skulls, however, and for this reason the Taung skull has come under intense scrutiny.

Much debate has focused on whether the Taung skull reveals that the australopithecine brain was reorganized along human lines from the brain of some ancestral ape. In our brain the expansion of the parietal lobe toward the rear has crowded the primary visual area into a small region at the very back of the brain. As a consequence the primary visual cortex constitutes only 3 percent of the human neocortex, compared to 15 percent in a monkey. Separating the parietal lobe from the primary visual area is the lunate sulcus. Anthropologists who have claimed to discern a replica of the lunate sulcus in the sedimentary filling of the Taung skull have not agreed on its position. Those who identify the sulcus in a posterior region conclude that the australopithecine brain had evolved toward the human configuration through expansion of the parietal lobe at the expense of the visual cortex. Those who locate the sulcus farther forward view the rear portion of the *Australopithecus* brain as being more apelike. The prominent South African anthropologist Phillip Tobias, who would appear to have no ax to grind, comes down decisively on the side of indeterminacy: it is not possible to locate the sulcus in the fossil, he has concluded. The issue is important, because substantial restructuring of the brain before it expanded much beyond the size for an ape would suggest that enlargement was a secondary phenomenon, that something remarkable preceded expansion of the brain. As I will explain below, such restructuring seems unlikely for theoretical reasons. Rather, restructuring appears to have been linked to expansion.

An adult male *Australopithecus* was nearly as large as a modern male chimpanzee, but its brain was only 10 or 15 percent more voluminous. Clearly, the brain of *Australopithecus* was a bit larger than that of an ape, and it appears also to have been slightly more human in configuration. Casts of fossil skulls reveal that the frontal and temporal regions were expanded slightly in the direction of modern humans. Furthermore Ralph Holloway of Columbia University has shown that Broca's area, which we now employ for speech, was slightly enlarged relative to its proportion in an ape. These traits of the australopithecine brain, as well as the slight overall enlargement beyond the level of an ape, suggest the presence of slightly greater mental capacity than that of a chimpanzee. Gestural communication was possibly more advanced than in apes, even if verbal expression remained without syntax comparable to that of humans. In any event, the relatively small size of the australopithecine brain points to conceptual powers much closer to those of apes than to those of modern humans, and the failure of *Australopithecus* to evolve an advanced toolmaking culture during its lengthy stay on earth would seem to confirm this inference.

The Growth of the Big Brain

I have discussed several ways in which the human brain differs in its proportions and functions from the brain of an ape. How did it change its configuration as it evolved its huge size? In general, primate brains that are large have relatively expanded volumes of neocortex. In a chimpanzee, for example, the neocortex forms a larger proportion of total brain volume than it does in the smaller brain of a monkey. The general trend for primates predicts that a brain as massive as ours should have an exceptionally large volume of neocortex, yet the human neocortex turns out to be even more voluminous than expected. Furthermore, the regions devoted primarily to thinking are especially large. These regions, known collectively as the association cortex, are the ones that are not dedicated to sensory or motor functions.

It appears that the great evolutionary expansion of the human brain automatically produced a disproportionately large volume of tissue devoted to thinking. Simply put, brain size greatly outstripped body size, and the excess brain tissue—the portion not required to operate the body—was available for higher functions. How the higher functions appropriated the brain tissue has only recently come to light, at least in principle. The basic concept is what Harry Jerison of the University of California at Los Angeles has termed the "principle of proper mass." According to this principle, the mass of neural tissue devoted to a particular function is appropriate to the amount of information processing that the function entails. In effect, as it grows, the brain organizes itself according to this principle.

Recall that the brain is fully supplied with neurons before birth—oversupplied, in fact, in that many neurons soon disappear, even though none are supplied to replace them. In both monkeys and humans, the phase of fetal brain growth in which neurons proliferate rampantly begins about forty days after conception. This phase lasts for about a hundred days in monkeys and about twenty-five days longer than this in humans. The proliferation occurs deep within the brain, and the myriad neurons assume particular positions in the neocortex by migrating to locations in this outer region that are specified by genes. In other words, the movements of neurons are genetically programmed. Through their migration the neurons build the six layers that constitute the neocortex, starting with the innermost layer and ending with the surficial one. The human neocortex is identifiable about two months after conception, and cell migration ceases by the end of the fifth month. Our neocortex assumes its wrinkled configuration as cells are added and crowd together. Another aspect of maturation is myelination, a process in which fatty sheaths encase neuronal fibers, insulating them and improving their ability to conduct electrical signals.

To some degree the connections that neurons make with one another are genetically programmed, but the genetic controls are imperfect and feedback from the body and its sensations influences both the production and the deterioration of particular connections. The neurons proliferate interconnecting fibers in the embryonic brain. Cells that form synapses—connections between neuronal fibers—receive more nutrition than those that do not, and those whose synapses fire off frequent messages are especially well supplied. Thus begins a remarkable warfare between neurons that belong to the same brain—a life-and-death competition in which those neurons that make many connections and receive stimulation survive, while those that do not, wither away. The kicking, arm flexing, and thumb sucking of a fetus, for example, stimulate synapses. Much of this kind of natural selection at the cellular level takes place in utero, but the process continues well beyond birth. A human baby enters the world with perhaps a trillion synapses connecting its cortical neurons, but a large fraction of these disappear during the first decade after birth.

Predictably, the maturation of brain tissue parallels the maturation of brain functions. The primary motor area, for example, matures ahead of the temporal lobe, which functions in human language control, just as we begin to put our arms and legs to use before we speak. On a finer scale, within the motor area of the brain, the region that governs the arms develops more rapidly than the region that controls the legs, just as movements of the arm come under control earlier than the movements of the legs: an infant can grasp objects before it can walk.

The development of brain and body entails reciprocal feedback systems. Maturation of a particular sector of the brain stimulates activity in a corresponding area of the body or in a connected part of the brain. The stimulated feature then matures more rapidly through use, but this very use promotes further development of the portion of the brain that activates it. As a human baby grows into a toddler and then a school-age child, the brain guides interactions with the environment. This relationship also molds the developing brain, however, favoring the fixation of beneficial neural maps and allowing useless neural connections to wither away. The result is that, although genes specify some traits of the developing brain, particular neuronal maps are forged through interaction with the environment, especially in the later stages of development. Patterns of bodily movement constitute one class of environmental interaction. Education constitutes another. Juvenile play enhances reflexes and coordination, which entail both mental and physical traits. It also develops the imagination.

The particulars of brain development remain to be uncovered, but Terrence Deacon has interwoven original observations with existing knowledge to produce a compelling general model of brain expansion that applies to the evolution of the huge brain of *Homo*. This model builds on Jerison's principle of proper mass. Various parts of the body, as they grow, send out fibers to the brain. The body automatically sends a total number of fibers to the brain that is proportional to its own—the baby's—size. This means that when the human brain, in growing so large, outstrips the body, there is no way the body can recruit the excess neurons and neu-

ronal connections that develop within the brain. More specifically, when the brain suddenly evolves to a much larger size without a corresponding expansion of the body, the sensory and motor
regions of the brain—and also the auditory and visual regions—
are at a disadvantage in competition for connections with new
brain cells. Neurons outside the brain are probably no more numerous than when evolution began to expand the brain. Areas of
the brain such as the prefrontal cortex are connected not to neurons of the rest of the body but only to neurons within other parts
of the brain itself. When the human brain originally expanded,
these regions were therefore favored; they had the right wiring and
received nourishment. In this manner areas of the brain that were
devoted to conceptual functions expanded dramatically, while
those concerned with bodily functions underwent little change.
Thus, as Deacon has pointed out, it is no accident that the prefrontal cortex of the human brain expanded preferentially, for this
is the region that plays the dominant role in thinking.

The expansion of regions of the neocortex devoted to thought
opened the way for the appearance of areas devoted to special
conceptual functions such as language. In Deacon's model these
areas differentiated out of preexisting areas, with which they retained some similarities. The new functions arose because axons
invaded new target areas of the brain, forming new neuronal connections.

Deacon's model squares with my own idea about the evolution of asymmetry in the human brain. If new areas had differentiated out of old areas in both hemispheres of the expanding brain,
they would have totally displaced essential preexisting functions.
The emergence of a new function in only one hemisphere, however, conserved the displaced function in the other hemisphere.
Overall brain expansion compensated for the disappearance of
the original function from one side by expanding the tissue devoted
to it on the other side.

This "displacement model" implies that heightened intellec-

tual powers were a natural consequence of the evolution of the
large brain of *Homo*. Because there was no commensurate increase in body size, the added brain tissue was available for thinking. A pattern observed for monkeys suggests that the displacement
model is on target. A few species of monkeys that have unusually
large brains for their bodies also have brains that resemble ours in
their general proportions.

   Having examined how the advanced structure and function of
the human brain are related to the organ's great volume, we confront a more fundamental issue: how and when did the brain of
our ancestors actually evolve to a large size? Did an early version
of the modern human brain, much bigger than the brain of *Australopithecus*, emerge with the evolutionary origin of *Homo*? Also,
how was the origin of the big brain related to the onset of the Ice
Age? The geologic record, I believe, has yielded answers to both
of these questions.

CHAPTER SEVEN

A Catastrophic Birth for *Homo*

Scalpels and scanning devices may allow us to peer inside the massive brain of *Homo sapiens* and see what evolution molded from the gray matter of *Australopithecus,* but there remains the issue of timing. Did evolution take a large step in the direction of modern humans early in the Ice Age when it brought forth early *Homo?* Even without detailed inquiry, a remarkable irony gives us reason to believe that early in the Ice Age *Homo* might indeed have emerged rapidly as an entirely new kind of big-brained animal. This is the paradox of our genus's having been born of a catastrophe. The wrenching contraction of the woodlands that had long offered *Australopithecus* safe haven above the ground presented an entirely new opportunity to any population of *Australopithecus* that straggled into the Ice Age—the chance to evolve a much larger brain.

Top-heavy Infants

The way in which *Homo sapiens* grows its large brain allows us to deduce that a population of *Australopithecus* could only have evolved such a structure after abandoning the habit of climbing trees every day. The human pattern of brain growth early in life is unique among primates; the key interval is the first year after we are born. During this brief period an average infant adds slightly more tissue to its brain than it will add throughout the remainder of its life!

Of course, a newborn baby also has a very big head for its body. The brain makes up slightly more than 10 percent of total body weight at birth. It may seem surprising that the same can be said for a newborn chimpanzee or monkey. In fact, this so-called 10 percent rule for brain weight at birth holds for primates generally. It is only during infancy that humans surge ahead of lower primates in brain size. After birth a monkey or chimpanzee fails to maintain the high rate of fetal brain growth that endowed it with such a large head when it entered the world. It embarks almost immediately on the second phase of growth—what I call Phase II—in which its brain expands much more slowly all the way to adulthood. Humans differ from lower primates in retaining the high fetal rate of brain growth—Phase I—through the first year of life after birth. The result is a one-year-old infant who is endowed with an enormous head that houses a brain more than twice as large as that of an adult chimp. Not until an age of about one year do humans settle into the sluggish Phase II of brain growth. Then, between our first birthday and adulthood, while we grow in total body weight by about 800 percent, our brain grows by only about 50 percent.

Smart but Immature Infants

Rapid brain growth is not an isolated feature of development in primates. It is linked to other aspects of maturation, which complicates matters considerably. In humans the persistence of the high fetal rate of brain growth beyond birth amounts to a retardation of the brain's development. It is not simply our brain that matures slowly, however, but our entire body. This condition arose because natural selection found no way of singling out the brain for delayed maturation. It accomplished the delay by slowing down the overall rate of bodily development immediately after birth. The result is that, although we grow rapidly in physical size after birth, we remain physically helpless while the fetal pattern of brain growth more than doubles our brain size by the time of our first birthday. Although our brain then switches abruptly from the Phase I to the Phase II pattern of growth, our slow overall rate of maturation lingers on. We continue to lag far behind apes in the level of physical development throughout our growing years.

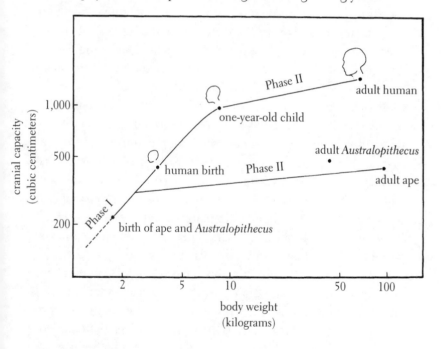

FIGURE 7.1

Patterns of brain growth. The pattern for *Australopithecus* resembled that of a chimpanzee or orangutan, whereas in humans the continuation of the high pre-natal rate through the first year after birth produces a much larger brain even at an age of one year, when the slower, Phase II, rate begins.

Many aspects of our sluggish maturation are quite obvious. Human females first menstruate at an average age of 13.7 years, whereas apes typically experience menarche before the age of 9. Human bones and teeth also mature much more slowly than those of apes. Adult humans can gain some sense of our prolonged schedule of tooth development by remembering that we flashed toothless smiles after we had learned to read. The schedule for apes is less protracted. Whereas human front teeth generally emerge between the ages of five and nine, for example, those of an ape move into place between the ages of four and a half and seven. Toward the rear of the mouth, our dental development proceeds even more slowly. Our wisdom teeth generally do not erupt until we are between fifteen and twenty-two years old, if they emerge at all, but those of an ape are in place between the ages of eight and twelve. The sutures between the bones of the human skull also remain open for a very long time—into the third decade of life—whereas those of an ape begin to join early in childhood. Our acquisition of motor skills is delayed as well. We are so helpless at birth that we cannot even crawl on our abdomen or pull ourselves into a sitting position before the age of ten months or so. Not until our first birthday are we likely to manage lumbering on all fours, like a bear, and only about three months later can we walk unassisted and feed ourselves. Obviously, infant chimpanzees and orangutans, which hang onto their climbing mothers at peril of death, are physically far more mature than human infants of the same age.

There can be little question that the two human features I have described, our prodigious brain growth immediately after birth and our infantile helplessness, are connected aspects of our maturation. One indication of this linkage is the simple fact that both of these traits represent severe developmental retardation. Our brain grows at the rapid fetal rate during our first year outside the womb, and our body in many ways remains fetal as well. It has been said that our true gestation time is not nine months but twenty-one—the in utero interval plus the ensuing twelve months of continued rapid

FIGURE 7.2

The resemblance between a human and a juvenile chimp. Our relatively flat face, weak brow, and tall forehead are among the juvenile traits of our ancestors that we retain into adulthood.

brain growth. The linkage between our dramatic postnatal brain growth and delayed maturation is especially evident in the fact that we hold the world's record in both categories. Of all mammals we exhibit the most dramatic increase in brain growth immediately after birth, and we also experience the longest interval of infantile helplessness. No other species of mammals more than doubles its brain size by its first birthday, and no other species requires fifteen months to begin to walk without parental support.

It might benefit modern humans if women had the size and strength to carry offspring in their wombs for twenty months and then to give birth to relatively mature infants. This arrangement would saddle parents with much less feeding and tending of off-spring. Such a long gestation period would be mechanically im-possible, however. Human babies are as large as they can be because of a basic limitation on the size of a woman's pelvis. While women, despite their shorter stature, have evolved a pelvis that on average is wider than a man's, any further broadening would impose undue mechanical problems for two-legged locomotion. If, then, we view the fetal period of human development as being drawn out over twenty-one months, the reason the last twelve months of this interval end up being spent outside the womb is that by about nine months after conception the head of the fetus becomes too large to fit through the mother's pelvis. Natural selection, under the influence of this constraint, established the modern human gestation period. In a sense we humans are all, of necessity, born prematurely.

An Apelike Pattern for Australopithecus

Did a baby *Australopithecus* grow with the delayed pattern of a modern human, or did it mature more rapidly, like an ape? We must address this issue in order to identify the particular evolutionary changes that marked the emergence of *Homo*. Because *Australopithecus* lacked our intellect, we might suspect that it also lacked our degree of infantile immaturity and followed instead a schedule of development more like that of a chimpanzee or orangutan. Robert Tague and Owen Lovejoy of Kent State University addressed this issue by reconstructing the obstetrics of Lucy, the well-preserved female skeleton that belongs to *Australopithecus afarensis*. They concluded that her pelvic inlet—the skeletal portal for the birth canal—would have allowed her to give birth to a baby no larger than a newborn male chimp or orang. Like all primates, Lucy's new baby would have resembled these newborn apes in having a brain that constituted about 10 percent of its total weight. We know that it then grew to nearly the adult size of a chimp or orang, with a brain only marginally larger. In other words, from birth to adulthood *Australopithecus* would have grown very much as an ape does. Growth of its more or less ape-sized

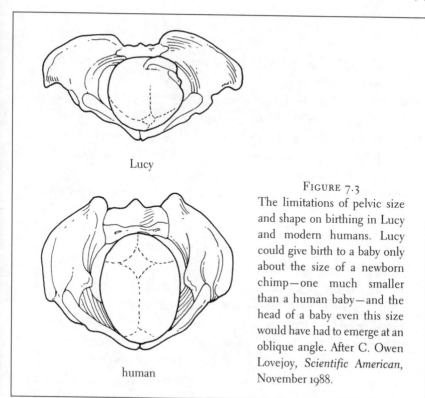

Lucy

human

FIGURE 7.3
The limitations of pelvic size and shape on birthing in Lucy and modern humans. Lucy could give birth to a baby only about the size of a newborn chimp—one much smaller than a human baby—and the head of a baby even this size would have had to emerge at an oblique angle. After C. Owen Lovejoy, *Scientific American*, November 1988.

brain would not have required prolongation of the high fetal rate of brain growth appreciably beyond birth, any more than this is required for the growth of an adult ape's brain.

If *Australopithecus* matured like an ape, then its newborn offspring would have been advanced enough to cling to their mothers almost immediately after birth. Given the upright posture of the mother, we might well imagine that the infant draped itself over her shoulders and wrapped its arms around her neck or clung to the long hair on the back of her head. Such an infant would not have obstructed the mother's ability to climb trees, an activity that I have contended was a necessary aspect of everyday life for *Australopithecus*.

How Human Was Early Homo?

Having contrasted patterns of development of *Australopithe-cus* and modern humans, we can similarly address the evolution-ary status of early *Homo* by measuring its fossil bones—if the requisite ones are available for study. What we need are estimates of four measurements for early *Homo*: we must know the brain size and body weight for a typical male both at birth and in adulthood. The numbers must be for a male because male babies are larger at birth, on average, than female babies and therefore tend more often to stretch a mother's birth canal to its limit.

The soft bones of babies are unlikely to be preserved as fossils, but we do not actually need to measure baby bones directly. Sur-prising as it may seem, we can derive all the numbers we need from fossil pelvises and skulls. If we know the pelvic inlet breadth for an adult female of early *Homo*, we can compare it to that of a mod-ern woman in order to estimate the average birth weight of its male offspring. It is then quite easy to estimate the brain weight for the newborn male—about 10 percent of total body weight, as it is for monkeys, apes, and humans. As for the body weight of an adult male, an empirical formula derives the number from the diame-ter of the hip socket. (The relationship holds because of the load-bearing function of the hip: the heavier the animal, the larger the ball-and-socket joint.) Finally, the interior dimensions of an adult fossil skull indicate cranial capacity.

We are fortunate that the fossil record has supplied the very bones needed to evaluate the developmental pattern for early *Homo*: a pelvic bone and a skull. These fossils appear to represent *Homo rudolfensis*, the earliest species of the human genus. One is the right half of an adult pelvis (labeled KNM-ER 3228). This well-preserved bone has a narrow sciatic notch near the tailbone, showing that it belonged to a male. From the large hip socket we can estimate that the owner of the pelvis was the size of a small modern man, weighing about 60 kilograms (132 pounds).

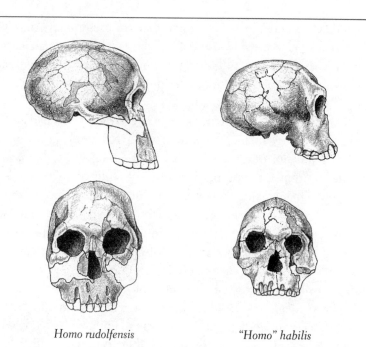

Homo rudolfensis   *"Homo" habilis*

FIGURE 7.4

Skulls of *Homo rudolfensis*, the earliest known member of the human genus, and the much smaller *"Homo" habilis*. The latter species, with its smaller brain and apelike limb proportions, does not deserve placement within the human genus.

To estimate the birth weight of a male *Homo rudolfensis*, we must follow a more devious path. We first turn the owner of the fossil pelvis into a female by adding 3 percent to the breadth of the pelvic inlet (the average difference between modern men and women). After his sex change, in our analysis the owner can stand in for the woman who bore him. Even this liberally augmented pelvis is much narrower than that of an average modern woman, however, and from the degree of difference we can estimate that an average male baby of the extinct species would have tipped the scales at only 2.1 kilograms (4.6 pounds) compared to 3.46 kilograms (7 pounds) for a Caucasian male in present-day America. Thus, the pelvis provides estimates of the weight of the male *Homo rudolfensis* both in adulthood and at birth. Less directly, with application of the 10 percent rule for newborn primates, it also gives the animal's brain weight at birth.

If the postnatal brain of the male *Homo rudolfensis* had grown like the brain of an ape, the animal would have ended up with an adult brain not much larger for its body than that of *Australopithecus*. In fact, the proportions of fossil skulls of unknown gender reveal that *Homo rudolfensis* had a much larger brain than this. An adult fossil skull of this species labeled KNM-ER 1470 has an estimated capacity of 760 cubic centimeters. This is quite large compared to estimates for various skulls of *Australopithecus*, which range from about 430 to 485 cubic centimeters. A portion of another fossil skull, KNM-ER 1590, points to an even larger brain size for *Homo rudolfensis*. Although this is only a large fragment of skull, we can see that it belonged to a very small child with a very large brain. A beautiful set of teeth in a jaw that accompanies the skull fragment reveals that the child was about six or seven years old. The skull's cranial capacity, though not precisely measurable, was about 800 cubic centimeters and would have approached 900 cubic centimeters in adulthood. While this adult figure is roughly 30 percent below the modern human average (1,220 cubic centimeters for women and 1,420 for men), it is nearly twice as large as the average estimate for skulls of *Australopithecus*. Clearly, in creating the brain of *Homo rudolfensis*, evolution moved profoundly in the direction of *Homo sapiens*, at least with respect to the volume of gray matter.

Even more to the point, the KNM-ER 1590 skull reveals that the brain of *Homo rudolfensis* expanded rapidly early in life. Even at the age of six or seven, its cranial capacity was more than three times what I estimate it to have been at birth. To grow its large brain, this young creature would have had to sustain its high fetal rate of brain growth while its body grew to just over twice its birth weight, just as humans do during the first year beyond the womb. It follows that *Homo rudolfensis*, the first species of the human family, experienced an interval of infantile helplessness that was comparable to our own. Although this conclusion is based on single measurements rather than abundant statistics, the results point to such a pronounced pattern of delayed development that they appear to be quite robust.

Homo rudolfensis: A New Kind of Animal

The near doubling of cranial capacity that occurred with the emergence of *Homo* from *Australopithecus* was a dramatic shift, but there is more to a brain than its volume. In fact, the brain of *Homo rudolfensis* exhibited a new shape. The cranium was higher and rounder than in *Australopithecus*, standing taller above the arched brow ridges that surmounted the eyes. Although we cannot assess the relative sizes of the major lobes of the brain precisely from fossil skulls of *Homo rudolfensis*, the elevation of the forehead from its low, sloping configuration in *Australopithecus* indicates expansion of the frontal lobes. The implications of this change are profound, because the frontal lobes are where humans process information and make logical decisions. As described in the previous chapter, it is here where we conduct the uniquely human function of planning for the future. Thus, it appears that the origin of *Homo* added human qualities of conceptualization to what had been a much more apelike brain.

It is enlightening to portray bodily features of *Homo rudolfensis* through comparisons with *Australopithecus*, its immediate ancestor. This approach reveals that the origin of the human genus entailed more biological changes than retardation of development and enlargement of the brain. For example, the brow ridges were less massive in *Homo rudolfensis* than in *Australopithecus*, reflecting weaker jaw muscles (which attach to the ridges). The jaws were also less projecting, which is to say, the face was flatter. Perhaps reflecting the same shift, the cheek teeth of *Homo rudolfensis* were smaller than those of *Australopithecus*, whereas, at the front of the mouth, the incisors were larger and more spadelike. These facial modifications, which were generally in the direction of modern humans but did not fully reach our condition, may reflect a change of diet that reduced the need for powerful grinding. The new animal presumably specialized less on coarse plant foods that the powerful jaws and broad molars of *Australopithecus* would have processed more effectively.

Below the neck *Homo rudolfensis* appears to have been quite human in form. The pelvis from which I estimated male birth weight for this species also gives an estimate of the animal's stature. From the diameter of the hip socket, Henry McHenry estimated that the creature who unintentionally donated his pelvic bone to science stood nearly 172 centimeters (5 feet 8 inches) tall. This is very close to the average for a modern man, but more than 21 centimeters above any of the eight estimates that have been made for individuals belonging to the genus *Australopithecus*. Two well-preserved thighbones (femora) of nearly the same antiquity as the KNM-ER 3228 pelvic bone are also assigned to *Homo rudolfensis*, and these turn out to be well within the size range for modern women. These two thighbones (KNM-ER 1472 and 1481a) and the pelvic bone are strikingly similar to equivalent bones of younger species of *Homo*. On the other hand, they contrast markedly with the corresponding bones of the stumpy *Australopithecus*. *Homo rudolfensis*, we can conclude, was a full-time ground dweller that approached modern humans in stature.

I have enumerated several features that *Homo rudolfensis* shared with modern humans. Those located below the waist included a deeper pelvis and longer legs than *Australopithecus*. Those above the neck included a brain that was not only larger than that of the ancestral genus but also expanded in the frontal region. *Homo rudolfensis* also had narrower cheek teeth, larger in-

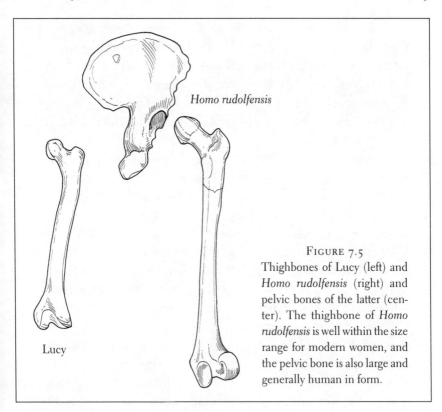

Homo rudolfensis

Lucy

FIGURE 7.5
Thighbones of Lucy (left) and *Homo rudolfensis* (right) and pelvic bones of the latter (center). The thighbone of *Homo rudolfensis* is well within the size range for modern women, and the pelvic bone is also large and generally human in form.

cisors, weaker jaws, and a flatter face. Finally, there was a general development delay that yielded a large brain not long after birth. We can view these traits as defining features of the genus *Homo*.

Bones and Tools of the Right Vintage

If the origin of *Homo* is to be attributed to environmental changes early in the Ice Age, it would be helpful to have tangible evidence that the genus was actually alive not long after 2.5 million years ago, when the Ice Age got into full swing. The fossil skulls and the pelvic and thigh bones from which I have inferred biological traits of *Homo rudolfensis* were all excavated from east of Lake Turkana (formerly called Lake Rudolf) in northern Kenya, and all date to the interval between 1.8 and 2 million years ago. Until recently we knew of no older remains that unequivocally belonged to this species. We did, however, have indirect evidence of the much earlier existence of *Homo*, some of which came in the form of stone tools.

Simple stone tools from Ethiopian deposits about 2.4 million years old came to light in the early 1980s. These were widely attributed to big-brained early *Homo*, but always with a degree of uncertainty. In general, Oldowan tools range from about 2.4 to 1.6 million years in age. For the most part they are simple flakes broken from larger stones. Most flakes were struck from the right sides of parent stones, suggesting that early *Homo*, like modern *Homo*, was predominantly right-handed. The flakes were not fashioned into refined shapes. At most they were given a bit of finish work—chipping of their margins, rarely on both sides, to give them a better edge. Presumably the primary function of the flakes was in food processing. The cores that remained after the flakes were struck may themselves have served in heavier duties. From the same strata as the manufactured tools come rounded stones, generally interpreted as hammer stones and anvils. These were perhaps used to crush nuts and other coarse plant foods.

For years the Ethiopian stone tools tantalized scientists by implying that *Homo* emerged near the start of the Ice Age, but doubts lingered in the absence of definitive bones. There were some claims that certain fossils—assorted teeth and one fragmentary temple region of a skull—represented *Homo*, but such assignments were controversial.

When I first published my new ideas about the origin of *Homo* in 1992, I was frustrated in having only the stone tools and controversial teeth and skull fragments as evidence that a big-brained member of the human genus was actually present very early in the Ice Age. Just a year later, a team led by the German anthropologist

Friedemann Schrenk announced a seminal find in the small country of Malawi, about a thousand miles to the south of the Kenyan collecting sites for early hominids. The new fossil was a well-preserved lower jaw of *Homo*, with many of its teeth intact, dating back to about 2.4 million years ago. It was the kind of discovery that I had been anxiously awaiting. Finally, in this single jawbone we almost certainly had the manufacturer of the oldest stone artifacts. It was *Homo rudolfensis*, from very early in the Ice Age.

The Terrestrial Imperative

We have now seen how the origin of the large brain of *Homo* was integrally linked to the physical retardation of infants, and we have learned that *Homo* was present very early in the Ice Age. Of course, the population of *Australopithecus* that evolved into *Homo rudolfensis* could only have done so after relinquishing its ancestral habit of tree climbing to the extent that it could tote and tend embryolike babies day after day, year after year. In other words, the transitional population must have conformed to the terrestrial imperative for growing a brain like that of *Homo*. It was presumably resigned to a life on the ground, where its forelimbs turned out to be unencumbered with the obligations of climbing, so that they were generally available for the parenting of feeble infants.

The key then to the evolution of *Homo* must have been a change in the way of life for the ancestral population of *Australopithecus*—the shift to a full-time life on the ground. Many years ago, Ernst Mayr, now a professor emeritus at Harvard, suggested that a change in the basic behavior within a group of animals has often preceded a major evolutionary shift of bodily proportions. Indeed, this generalization probably holds, because behavioral changes, even genetically programmed ones, tend to be easier for evolution to achieve than wholesale anatomical restructuring. The change of behavior that appears to have triggered the origin of

Homo was especially simple in that it was not a shift to a totally new lifestyle but only the abandonment of one segment of a previously broad behavioral repertoire. A life divided between terrestrial and arboreal activities contracted to become a life lived fully on the ground. For the origin of *Homo*, of course, it is more than mere conjecture that behavior changed before body and brain. This ordering was essential for the anatomical evolution to occur; it was dictated by the terrestrial imperative. To envision how the total sequence of changes came about, we must return to the topic of environmental change in Africa early in the Ice Age.

A Catastrophic, Punctuational Birth

Although the contraction of lifestyle for the population of *Australopithecus* that evolved into *Homo* was an easy shift from the standpoint of behavior, it must have constituted an ecological crisis. The population obviously survived the vicissitudes of the early Ice Age, but its life must have been quite brutal. *Homo* faced hardships similar to those that were exterminating its australopithecine kin. In particular, frequent attacks by carnivores must have decimated its ranks. We will never be able to reconstruct the travails of the transitional population or track its evolutionary course in detail. Delineation of its geographic boundaries for even a brief moment in time is out of the question; its fossils are too rare and our means of dating them, too imperfect. But we do have evidence that this pivotal population had evolved into *Homo* before the last of its close austalopithecine relatives had vanished.

In fact, we have evidence that *Homo* may have emerged from *Australopithecus* by way of a punctuational branching event. The discovery of the very early UR 501 jawbone far to the south of most other East African fossil homid sites raises the possibility that *Homo rudolfensis* evolved in this region or even farther south from a population of the southern species, *Australopithecus africanus*. Recall that the Taung skull of this species may be as young as 2.6 or 2.4 million years old. A future shift of this estimate to the younger extreme of the present range would increase the likelihood that then *Australopithecus africanus* overlapped in time with *Homo rudolfensis*, whose jawbone also dates back to about 2.4 million years ago. New reports of stone tools slightly older than 2.5 million years suggest an even earlier origin for *Homo*. The ancestral genus had existed with little change for a million and a half years or more, and an overlap with *Homo* would point to evolutionary branching, or speciation.

Two things suggest that this speciation event would have been punctuational, or abruptly divergent. One is the lack of fossils that are intermediate in form between those of *Australopithecus* and those of *Homo*. The other is the apparent absence of stone tools more primitive than those of the Oldowan culture. The fact that the earliest bones of *Homo rudolfensis* and the earliest tools attributed to it are of similar age suggests that the new creature wasted little time in putting its huge brain to good use. The absence of recognized intermediate bones does not mean that there never were any, but it does mean that any intermediate forms probably existed during only a brief interval of geologic time and in small numbers within a restricted geographic area.

How might *Homo* have originated rapidly if not by way of punctuational speciation? The alternative is what I term bottlenecking, a process that does not entail branching but instead amounts to the evolutionary rebirth of a species. When the environment turned against the species of *Australopithecus* that gave rise to *Homo*, the ancestral species may have shrunk to become a small population confined to a narrow geographic area. From this bottleneck *Homo* may have emerged by the evolutionary transformation of the beleaguered population. Thus, *Homo* would never have overlapped in time with the ancestral species of *Australopithecus*, whereas if *Homo* arose by way of a speciation event the two would have coexisted, at least briefly.

It is sobering to recognize that *Homo* might well never have emerged from the catastrophic changes in Africa early in the Ice Age. The *Australopithecus* population that made the evolutionary leap could easily have gone the way of the others that shrank to oblivion. If the group that turned into *Homo* was quite small compared to the total population of *Australopithecus* alive at the dawn of the Ice Age, then devastation was the norm for troops of the ape-like creature, and it was only by the slimmest of chances that *Homo* ever came into being.

This is not to say that we cannot imagine why natural selection would have created the human genus by favoring a large brain. The point is that many troops of the ancestral species probably failed to undergo such natural selection. Presumably by chance, the requisite changes—mutations or recombinations of genes—failed to occur within these populations before they dwindled to extinction under the new environmental stresses. Within these populations, then, natural selection never had the raw material from which to fashion a brain like that of *Homo*.

A Test

We cannot test directly the idea of a catastrophic birth for *Homo*, but we can put the idea to an indirect test, which it turns out to pass with high marks. The test is based on a simple prediction: if, indeed, the drastic reduction of forests early in the Ice Age triggered the transformation of the human family, then other groups of African mammals that depended on forests for survival should have experienced similar changes. Henry Wesselman of California State University has shown that the so-called micro-mammal fauna, which includes rodents and other small forms, did indeed undergo such a shift in southern Ethiopia. Species that were adapted to moist conditions gave way to species, such as gerbils, that occupy arid habitats. Unfortunately, we remain ignorant of the history of small mammals elsewhere in Africa, but thanks to the work of Elisabeth Vrba, we know that antelopes experienced this kind of change throughout the continent.

Though imperfect, the dating of antelope fossils is adequate to show that, from Ethiopia to South Africa, many species of forest-dwelling antelopes died out about two and a half million years ago and that within a hundred thousand years or two a variety of species adapted to grassy habitats appeared. Many of these new types survived to roam the savannas and open woodlands of modern Africa.

Having viewed the history of antelopes as a test of the idea that changes in vegetation influenced human evolution, we can also turn the argument around. First, let us swallow our pride and acknowledge that the immediate ancestors of our genus were basically large mammals. We can then ask why the onset of the Ice Age should not have affected these creatures as it did other large mammals. Forest-dwelling antelopes suffered heavy extinction, and during the same general interval natural selection produced species of antelopes adapted to the new conditions. Is it, then, surprising that *Australopithecus* should also have died out and in the process given rise to a new genus of hominids that could cope with the more open habitats?

A Different Tack

The general idea that the Ice Age changed the course of human evolution is not new. Vrba has offered several hypotheses to this effect, including the idea that the spread of grassy terrain might have caused hominids to expand their feeding habits or to evolve better abilities for locomotion on the ground. She has also proposed that colder climates might have caused the evolution of the delayed maturation that characterizes *Homo*, perhaps by affecting the operation of certain hormones. This is a difficult idea to embrace. First, although evolution in cold climates may have delayed the development of certain organs within some groups of insects and salamanders, there is no reason to believe that what happened to a few members of these lowly groups should have happened to human ancestors. Second, although climates have cooled slightly in equatorial Africa during the modern Ice Age, they have remained tropical.

The idea of a catastrophic birth for *Homo* follows a different line of reasoning. Paradoxically, this idea invokes an environmental crisis as the essential catalyst for the evolution of the human genus. The idea gains credibility by also explaining the evolutionary stagnation of *Australopithecus* up until the time of the crisis by showing how no population of this apish form could have evolved into *Homo* until a shrinking of woodlands forced its members to satisfy the terrestrial imperative by abandoning their dependence on trees. The environmental crisis that compelled populations of *Australopithecus* to live fully on the ground did not actually force *Homo* to evolve but simply opened the way. It propelled a population of *Australopithecus* across an evolutionary barrier, on the other side of which lay vast new opportunities for natural selection to exploit.

The Great Evolutionary Compromise

While it is possible to imagine many ways in which the large brain of *Homo* would have been useful, its evolution also entailed a highly detrimental side effect—singularly immature infants. Parents now had to feed and protect not only themselves but also their feeble, virtually immobile offspring. Some mothers would have had more than one needy child in tow. These and other special problems of child rearing must have beset members of the human genus since its inception.

Conspicuous as the problems imposed by helpless infants may

be, even for modern humans, some prominent scientists have downplayed them, assuming that because the retardation of early human development is so profound it must actually have redeeming value. These scientists have then been at pains to explain why helplessness and dependency early in life should be beneficial simply because it evolved. In a book entitled *Growing Young*, the anthropologist Ashley Montagu enumerated several ostensibly valuable attributes of delayed human maturation. Among these, he cited the provision of a long interval for learning from elders, for bonding to a family unit, and for remaining cooperative in a complex society.

The fact is that we learn very little during our first two years of life and cooperate largely through unwilled passivity. It certainly would benefit parents if our bodies could mature more quickly during this early interval. Furthermore, it is not necessary that we remain physically immature in order to remain attached to older people as dependent trainees. As college students prove, we remain more or less tractable and educable long after we reach physical maturity; in other words, we often choose to remain socially—and, in effect, ecologically—dependent long after we have the bodies of adults. In general, in the emotional and cultural aspects of our lives, we cannot be deriving a net benefit from retarded physical development.

The evolution of the large brain of *Homo* by means of a problematical slowing of development amounted to a profound trade-off. It must stand as one of the most remarkable evolutionary compromises in the history of life. On the negative side were the physical and mental deficiencies of immature offspring that, from the earliest days of *Homo*, constituted a great ecological handicap for parents. From the beginning extended child rearing has robbed parents of time that they could otherwise have spent gathering food, making tools, or constructing shelters, and it has restricted mobility and complicated confrontations with enemies. On the positive side were the vast benefits of the new brain. In the game of natural selection, the positive value of the large brain clearly out-weighed the negative side effects of infantile immaturity. Other-wise, quite simply, our brain would never have evolved.

For natural selection to create the large brain of *Homo*, the many benefits conferred by the incipient brain not only had to out-weigh the problems imposed by helpless infants but also those im-posed by the high rate of metabolism of the large brain itself. Recall that brain tissue requires an enormous supply of energy. Fatty meat and bone marrow are rich sources of energy, and, as Robert Martin of the Anthropologiocal Institute of Zurich has ob-served, we can imagine that early *Homo* turned to them increas-ingly as its brain evolved toward larger size. The brain itself would have played an important role in the capture of animals that sup-plied the meat and marrow. In other words, the large brain of early *Homo* must have played an important role in stoking its own metabolic furnace.

One might wonder why evolution failed to follow some differ-ent path to the large brain that would have entailed no great eco-logical sacrifice. In fact, all such paths were probably highly improbable. A delay of the developmental process was probably the only mechanism by which evolution could have readily produced the marked encephalization. By employing this mechanism, nat-ural selection made use of a pattern of growth—the high rate of brain growth in utero—that was already present in the ancestral an-imal. All that was required was a change in timing. This mode of evolution also had the advantage of producing a big brain early in life, so that advanced learning could begin during childhood.

Why Evolve an Enormous Brain?

How exactly might a very large brain have improved the chances of survival for the incipient *Homo* if it required so many calories and burdened its owner with highly immature infants? One way to begin to answer this question is to ask another. What basic mental traits separate modern humans from lower animals and what might those traits have accomplished for early *Homo?* The traits that require large frontal lobes of the brain presumably began to appear during the emergence of *Homo.* We have seen that generative, or creative, ability is an especially salient trait of this kind. From it, early on, should have come the potential to fabricate tools, including weapons that could have warded off carnivores that were swifter, stronger, and innately better armed than members of the human genus. In fact, I suggest that the need for self-defense while living freely on the ground was the primary driving force behind the natural selection that created the large brain of *Homo.* This is not to say that the brain had little value in the quest for food—a quest that may have become more difficult once the transitional population was living fully on the ground. Having access to fewer of the resources that grew in trees, this group probably faced greater food shortages than *Australopithecus* had normally encountered. As we can see from the history of our own species, at some point tools facilitated the acquisition and processing of both meat and edible plant materials on the ground.

From the new generativity would also have come the capacity to construct a language, perhaps largely through signing at first. Early *Homo,* using its newly expanded frontal lobes, was probably the first creature on earth that could generalize from experiences of its past in order to devise strategies for the future. As brainpower advanced, societies would have used the experiences of earlier generations to mold their strategies, because advanced communication gave rise to an oral tradition. By combining its ability to communicate with unprecedented creativity, early *Homo* presumably engaged in social cooperation that was not rigidly instinctive, like that of its four-legged enemies, but unconstrained and versatile.

Pair-bonding

When they engaged in hunting or self-defense, only cooperation with one another would have enabled the males in a troop of
early *Homo* to substitute brains for the brawn they lacked. This
need for rapport among males probably accounts for the origin of
pair-bonding, which is the semipermanent attachment of one
male to one female. In species of primates that lack such bonding,
males compete intensively for females. Sometimes they actually
fight, but the danger of injury to both opponents has led evolution
to produce substitute behavior, in which potential combatants
simply face off in aggressive postures, with one eventually intimi-
dating the other. The result is what Darwin labeled sexual selection — the process described earlier in which members of one
gender that win contests for mating rights produce a disproportionate number of offspring.

The "height-is-might" principle comes into play in sexual selection. More often than not, the larger of two contesting males is
the more imposing and wins the right to mate with the one or more
females who are the object of competition. As a result, in species
where promiscuity rather than pair-bonding is the rule, males are
usually much larger than females. As generation has followed generation in the evolution of such species, the genes of big males have
accumulated preferentially. Living primates that mate promiscu-
ously illustrate the results of such sexual selection for large males.

In most primate species that practice pair-bonding, males are
closer to females in size. The current American divorce rate
notwithstanding, pair-bonding is an innate pattern for *Homo sapiens*, and on average modern men are only slightly more than 25
percent heavier than modern women. The fact that men are larger
at all is probably related to their early role in hunting and, perhaps,
warfare. On average women are probably just big enough to give
birth to somewhat larger than average male babies.

According to the criterion of body size, *Australopithecus*

afarensis—the species to which Lucy belonged—probably did not engage in pair-bonding. Lucy weighed only about 30 kilograms (66 pounds), but other skeletons assigned to her species represent animals that were nearly twice as heavy and are presumed to have been males. The disparity seems to have been smaller for *Australopithecus africanus*, the younger species that was probably ancestral to *Homo*. Even within this southern species, however, males may have been as much as 70 percent heavier than females. At the *Australopithecus* stage of our family history, from more than 4 to about 2.3 million years ago, promiscuous mating apparently was the norm.

Human ancestors may have had little reason to engage in pair-bonding until members of a troop were cooperating in complex ways. Males could have worked more compatibly within hunting parties if, instead of vying with each other for females on the home front, each understood that a particular mate awaited him on return from a hunting expedition.

It is also easy to see how pair-bonding would have benefited *Homo* in rearing its physically immature offspring. Natural selection probably favored any male who became part of a nuclear family in order to help train his own offspring. These progeny were more likely to survive and reproduce, passing on their father's genes, than were offspring that a father left in the care of an unaccompanied mother during their lengthy childhood. Likewise, natural selection probably favored females who were inclined to enter common-law marriages, which favored dependent children with two devoted parents.

How Human was Homo rudolfensis?

Oldowan tools and fractured bones associated with them point to a major shift from the australopithecine pattern of subsistence at the time *Homo* came into being. Cut marks on bones that have been found with manufactured stone flakes reveal that the makers of the tools—presumably populations of *Homo rudolfensis*—butchered large animals, especially antelopes. The toolmakers evidently put sharp flakes to use in skinning the animals and in cutting through their tendons and muscles to separate flesh from bone. Some of the bones also bear the unmistakable tooth marks of carnivores. In some fashion human ancestors and four-legged meat eaters were sharing carcasses, but it is not clear which group usually got there first. Which ordinarily was the killer and which, the scavenger? Attempts to resolve this question have stalled in the absence of definitive evidence. If the troops of *Homo rudolfensis* were the killers, they were big-game hunters. If, instead, they were habitual scavengers, they must have had the wherewithal to intimidate other large flesh eaters. In either case they were engaging in effective aggressive activity that they could only have undertaken by pooling both their prodigious mental resources and their limited physical powers to best animals that were swifter and stronger than they. Their butcher sites reveal that they were also collaborating in order to process meat. The rendered meat presumably formed part of an omnivorous diet much like that of hunting and gathering cultures of our own species.

Given the evidence of cooperative carnivorism for *Homo rudolfensis*, we can imagine how pair-bonding would have benefited this species by reducing disruptive rivalry between males. We also have seen how pair-bonding would have aided the species in rearing its immature infants. It would be satisfying to find evidence for pair-bonding within *Homo rudolfensis* in the relative body sizes of its two genders. Unfortunately, although we have some indications here, the data are thin. The mature male pelvis from which I estimated the size of a male of the species, both at birth and in adulthood, has a hip socket that would have housed a femoral head (the ball at the upper end of the thighbone) about 47 millimeters across; this happens to be only about 2 millimeters smaller than the average diameter for adult modern American males. The KNM-ER 1481a thighbone, which is of nearly the same vintage as the pelvis, has a well-preserved femoral head that is just a shade larger that the average for adult modern American females. It is tempting to conclude that the fossil pelvis belonged to a male and the fossil thighbone, to a female—and then to assert that *Homo rudolfensis* resembled us in the relative sizes of its genders. This, however, would be going too far. Within each gender of any species of the human family, there has been enough variation of thighbone proportions to allow a modest possibility that the fossil femur, like the pelvis, belonged to a male—simply a much smaller one. Only future discoveries can reveal whether *Homo rudolfensis* fits the typical body-size pattern for pair-bonding species. If I were to bet, it would be for the presence of pair-bonding in this advanced creature, a product of a giant evolutionary step toward the modern human animal.

The Ultimate Cause: An Old Idea

Any thorough discussion of the consequences of the Ice Age for human evolution must lead to a fundamental question: What triggered the expansion of continental ice sheets to begin with? What initiated the cycle of glacial expansions and contractions that has continued to the present day? This intriguing question resonates more strongly when we connect the onset of the Ice Age to the origin of our own genus.

Remarkable as it may seem, a popular idea as to what triggered the Ice Age has focused on a skinny neck of land that sits much closer to the equator than to the North Pole. This is the Isthmus of Panama, which rose out of the ocean slightly before three million years ago. The newly formed isthmus blocked the equatorial current that flows westward across the Atlantic. Before the barrier rose up, some of this current's tropical waters had passed between the Americas into the Pacific. The uplift of the isthmus deflected most of these waters northward, augmenting the Gulf Stream, which warms southeastern North America before angling across the North Atlantic toward the British Isles. It has been thought that the strengthening of the warm Gulf Stream might have caused glaciers to form adjacent to eastern North America slightly before three million years ago. The reason is that the Gulf Stream, like a hot tub on a cold morning, sends misty vapor into the air above the northern Atlantic. This moisture creates clouds, which in winter drop snow on eastern Canada, Greenland, and Scandinavia. During geologic intervals when the Northern Hemisphere is cool enough, as dictated by the earth's orbital dynamics, the winter snow piles up deeply enough to form glaciers in all of these regions.

Perhaps, then, snow first accumulated to glacial depths only after the Isthmus of Panama rose up to nourish the Gulf Stream. The problem with this idea is that we still live within the Ice Age, and the Arctic remains frigid even at this time of relative glacial recession. The frozen condition of the north polar region year-round cannot result from the presence of the lone Greenland ice cap and scattered mountain glaciers. Such a modest array of glaciers cannot be the cause, but must be a result, of a general cooling of the polar climate that persists even during a glacial minimum such as the present. We must view the general frigidity of the north polar region as the central feature of the Ice Age. In fact, before the Ice Age the north polar climate was quite mild. As described in chapter 4, the Arctic Ocean had been warm enough shortly before the Ice Age began that many species of Pacific mollusks adapted to temperate conditions were able to make their way through it to the Atlantic. Why did the big freeze-up suddenly occur?

The Ultimate Cause: A New Idea

While writing this book, I have found what I believe is the answer to the riddle of the Ice Age. For some years it had seemed that there might be a simple answer: the concentration of carbon dioxide in the atmosphere might for some reason have declined, cooling the entire earth and plunging the north polar region into unyielding winter. At a meeting of the American Geophysical Union in June 1995, Maureen Raymo of the Massachusetts Institute of Technology debunked this idea with convincing evidence on the isotopic composition of carbon in the marine plankton that lived just before the Ice Age. It turns out that these organisms floated in the ocean beneath an atmosphere in which the concentration of carbon dioxide was quite close to that of the present. In other words, the greenhouse effect had not been significantly weaker during the Ice Age than it was during the interval just before, when the the north polar region was temperate.

Before this new evidence emerged, the possibility of a greenhouse control had discouraged me from thinking very deeply about the puzzle of the Ice Age. Why spend time trying to solve a problem by looking at particulars of the ocean and atmosphere when a simple change in our planet's greenhouse seemed a likely, if untested, possibility? When I learned that this mechanism had failed a test—actually a simple measurement, albeit an indirect one—I put on my thinking cap, read a few relevant articles, and within a few days convinced myself that I had the answer.

This answer ultimately brings us back to the Isthmus of Panama—but not to the idea that its origin was a direct cause of continental glaciation. Rather, the key role of the isthmus was in changing circulation patterns within the world's great oceans and leaving the Arctic Ocean without a heat supply.

In addressing the riddle of the Ice Age, we must first ask what keeps the north polar region cold today. Then we must try to determine what change put this refrigeration system in place slightly before two and a half million years ago. This will have been the change that triggered the Ice Age.

When it comes to large-scale changes in patterns of heat transport over the surface of the earth, we must look to the oceans.

Water has a high heat capacity, meaning that it takes a great deal of heat to warm it. A result of this property is that water heats and cools air more effectively than air heats and cools water. In short, the oceans move a great deal of heat around. When oceans from low latitudes bathe a polar region in temperate waters, air moving over these temperate waters tends to become temperate as well.

Today, however, there is very little movement of temperate water into the Arctic Ocean. This polar sea is largely isolated from the oceans to the south. The Bering Strait provides only a very slender connection with the Pacific. The much broader gap between Greenland and Scandinavia offers space for more extensive exchange between the Atlantic and Arctic oceans, yet very little exchange now takes place. The reason for this separation is that a great ribbon of moving water robs the Arctic Ocean of heat from the Atlantic.

This influential ribbon of water, also known as a conveyor belt, flows from the central Pacific all the way to the margin of the Arctic Ocean and back again. It flows from the Pacific as a shallow current through the Indian Ocean, around Africa, and diagonally westward across the Atlantic, where it joins the Gulf Stream on a curved path northward from the Straits of Florida to the northernmost Atlantic. There, most of the composite mass of water descends and turns back to the south, passing at great depth through the Atlantic and around the tip of Africa to the Pacific. It then rises in the Pacific, north of the equator, and loops back, retracing its path to the North Atlantic as a shallow current.

The main driving force for the conveyor belt is the descent of its waters adjacent to the Arctic Ocean. The water that sinks there becomes what is known as North Atlantic Deep Water—water that flows back to the Pacific as the deep segment of the conveyor belt. The water sinks in the North Atlantic because it is denser than the surrounding waters. It is dense because it is not only cold but also a bit saltier than average seawater. The saltiness comes from its passage across the Atlantic through the zone of the dry trade

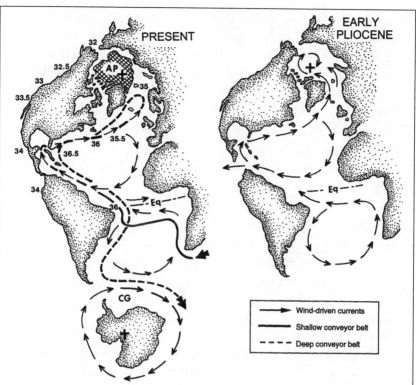

FIGURE 7.6

The role of the transoceanic conveyor belt. Today, the conveyor belt carries shallow water from the Pacific across the equator (Eq) to the northern Atlantic, where it sinks and then flows back to the Pacific at depth. The conveyor belt is driven by the great density of North Atlantic waters that results from their high salt content (numbers indicate salt content of water in parts per thousand). The descent of the conveyor belt prevents warm Atlantic waters from flowing into the Arctic Ocean, most of which is consequently covered by sea ice year-round (hachured pattern). The Arctic Pond (AP) that is thus formed, like the circumantarctic gyre (CG) of the south polar region, remains cold because of the long residence time of its waters. Early in the Pliocene Epoch, before the Isthmus of Panama formed, exchange of waters with the Pacific would have lowered the salt content of the Atlantic. Then, instead of descending, waters that flowed northward from the Gulf Stream would have continued into the Arctic Ocean, warming its waters and also the polar air masses.

winds, which sweep westward on either side of the equator. These thirsty air masses evaporate surface waters of the conveyor belt, concentrating their salt. The trade winds then cross the narrow continental region of Central America and convey much of their newly acquired moisture to the Pacific Ocean as rainfall. The resulting concentration of the Atlantic waters and slight dilution of the larger Pacific Ocean leave the Atlantic slightly more saline, on average, than the Pacific.

Of particular importance is the saltiness of the waters that reach the northernmost Atlantic. As these waters approach this region, they lose heat to the atmosphere, and the winds that pick up their heat carry much of it to northern Europe, keeping this region unusually warm year-round for its high latitudinal position. The ocean currents, already dense from evaporation, become even denser as they cool, and they then sink between Greenland and Scandinavia, driving the conveyor belt; the descending waters pull water northward through the Atlantic and push southward at depth, winding their way back to the Pacific.

The descent of the great conveyor belt deprives the modern polar region of Atlantic waters that would otherwise warm it by continuing northward into the Arctic Ocean. If we can explain why the conveyor belt should have come into existence at the time when the Ice Age began, then we will have an explanation for the origin of the Ice Age.

The oldest sedimentary deposits in Panama that contain fossils of animals adapted to nearshore marine environments date to between three and three and a half million years ago. This, then, was when a hump of seafloor rose up to form the Isthmus of Panama. The movement of the earth's crust that created the isthmus was part of a regional uplift that also added height to the nearby Andes Mountains.

Isotopes from fossil plankton in cores of deep-sea sediment reveal that the Ice Age also got under way slightly before 3 million years ago, with successively larger glacial maxima occurring between about 3.1 and 2.5 million years ago. By the latter date the Ice Age was in full swing. So the timing was right, but what was the connection between the isthmus and the conveyor belt?

In fact, the conveyor belt should not have existed before the isthmus formed. The reasoning here is quite simple. An open strait between the Americas would have allowed most of the westward flowing equatorial current of the Atlantic, with its unusually salty waters, to flow straight across into the Pacific. This flow would have more or less equalized the saltiness of the two oceans. As a result there would have been no unusually salty current flowing to the northernmost Atlantic and sinking to form North Atlantic Deep Water—no conveyor belt. There would have been only a weak submergence of cold water, as happens today in the northernmost Pacific, where the water becomes as cold as that of the northernmost Atlantic but is not salty enough to sink on a large scale.

In the absence of the conveyor belt, the Atlantic waters flowing northward should have continued their journey past Greenland and Scandinavia into the Arctic Ocean. They would then have spiraled clockwise under the influence of the Earth's rotation and returned to the Atlantic. The flow of temperate water through the northern ocean would have kept it relatively warm.

It is true that the northward flow of water in the Atlantic would have been weaker before the isthmus formed. In the absence of a conveyor belt, only the wind-driven Gulf Stream would have moved water northward. Furthermore, because the temperature gradient between the equator and the poles would have been more gentle than today, the winds driving the Gulf Stream would have been weaker. The Arctic is a small ocean, however, and only a small flux of temperate water through it would have kept it much warmer than it is today. Furthermore, today the icy polar region sends cold air masses down into the North Atlantic, cooling its waters. Before the Ice Age the waters of the Gulf Stream would have undergone less cooling as they approached the Arctic.

When the isthmus formed—and with it the great conveyor belt—everything changed. The Arctic Ocean, deprived of its source of heat from the south, became what amounted to a frigid pond. It simply assumed the cold temperatures that will characterize any body of water that is largely isolated and positioned at a very high latitude, where the sun's rays are spread thin as they strike the earth's surface at a low angle. Pack ice soon covered the North Pole and, in fact, a large portion of the Arctic pond. The ice reflected most of the sun's rays, causing the region to cool further. Across areas of nearby land, summers soon remained too cool to melt snow, and where storms were heavy enough, the snow accumulated to glacial thickness. Under the influence of periodic changes in the earth's rotational movements, the amount of sunlight reaching the planet varied through time, and the ice sheets waxed and waned to produce glacial maxima and minima. This, however, is a subplot, or actually an epilogue, because the climax of the saga was the initial plunge of the earth into the modern Ice Age.

Since the very start of the Ice Age, the north polar region has sent some of its frigidity far to the south. Even during the present glacial minimum, arctic blasts—cold high-pressure systems—from time to time spin down to bring winter chills to marginally tropical regions such as southern Florida. And as we have already seen, waters that well up from great ocean depths are cooler than they were before the Ice Age got under way, and, evaporating more slowly, they yield less moisture to regions such as equatorial Africa. In effect, low latitudes have been exporting little warmth to the Arctic pond, and the Arctic pond has been retaliating by sending very cold air and water southward, robbing both the land and sea of their customary heat, especially in winter. Because of colder winters, climates at midlatitudes are much more seasonal than they were before the Ice Age.

There is a satisfying parallel between the origin of the modern Ice Age in the north and the origin of the glacial interval in

Antarctica thirty-four million years ago, as described early in chapter 4. Like the northern event, the earlier one in the south entailed the entrapment and cooling of water close to one of the poles as a result of changes in ocean circulation brought about by movements of the earth's crust. The southern trap is a current that cycles round and round the pole, and the northern one is a polar pond. Both events also brought seasonally colder and drier climates to many regions far from the poles. It was partly because the Antarctic event came first that it was more wrenching to life on earth, causing a major global extinction. It wrought its effects on a world that, from pole to pole, had been bathed in warm, equable climates. The Arctic event, in contrast, descended on a world that the Antarctic event had already moved far down the path along which polar cooling could take it. The earlier event had already refrigerated the deep sea and the waters that welled up from it throughout the world; even the north polar region was cooler than it had been before. Given these changes, many forms of life were tolerant of seasonal climatic change by the time of the northern Ice Age.

From our human vantage point the second climatic event will never appear minor, however. Its ultimate cause, the formation of the Isthmus of Panama, was even smaller in scale, and yet it looms so large in our history that to contemplate its role here is quite humbling. Our genus, *Homo*, came into being fortuitously—only because, thousands of miles from its African birthplace, a narrow strip of the earth's crust happened to rise above the surface of the sea.

CHAPTER EIGHT

The Saga of *Homo*

He died facedown among the reeds. In the flooding waters turtles and catfish nibbled at his weakening flesh. When only his bones remained, hippopotamuses and giraffes inadvertently trampled them, fracturing a few, until the floodwaters returned so many times to spread their sand and mud that even his bulbous skull was buried. So ended the natural interment—as reconstructed by my former Johns Hopkins colleague Alan Walker, now at Pennsylvania State University—of the so-called Turkana boy. His burial site was on a floodplain of the ancestral Omo River in Kenya, just west of modern Lake Turkana. Walker and Richard Leakey have edited a book that puts some of the flesh back on the Turkana boy's skeleton, the most complete early fossil skeleton of the human family—more complete even than the collected remains of Lucy.

Introducing Homo erectus

The Turkana boy belonged to the species *Homo erectus*. Discovered in 1984, his skeleton reveals the look of our ancestors nearly a million years after *Homo rudolfensis*, the oldest known member of the human genus, came into being. In the boy's bones we see the results of the first step that evolution took within the genus *Homo*. The estimate of 1.6 million years for the boy's age is considered accurate because beds slightly above and slightly below the level of his resting place have yielded reliable radiometric dates. As it turns out, the boy and other members of *Homo erectus* resembled *Homo rudolfensis* in many ways and were almost certainly its direct descendant. The more recent species had a bigger brain, however, and the results are seen in marked cultural advances and possibly also in a new wanderlust: rather quickly after it came into being, *Homo erectus* became the first species of the human family to make its way out of Africa. In fact, it ambled all the way to the Far East.

The cause of the Turkana boy's preadolescent death will never be known. Walker's osteological autopsy has produced but a single hypothesis for his demise. The right side of his lower jaw bears a lesion alongside two of the teeth. Rather than coming out cleanly, the milk teeth had broken off not long before the boy died, leaving two bits of root behind. These remnants caused the adult premolar to rotate abnormally as it moved upward into its adult position, creating an abscess. It is conceivable that fatal blood poisoning was the result. Even so, myriad alternative causes of death, undetectable in the fossil bones, cannot be ruled out. What is clear is that predatory animals were not responsible for the boy's demise, because his bones bear no marks of gnawing. A watery grave, shallow though it may have been, probably sequestered the corpse even from terrestrial scavengers, much to the benefit of modern science.

The left and right sides of the boy's lower jaw were the first of his bones to be spotted partly entombed in the substratum. (Loose bone fragments had turned up earlier.) His jawbones were embedded in the base of an acacia tree, for which the boy's upside-down braincase had apparently served as a subterranean flowerpot. His skull is remarkable in retaining nearly all of its teeth, although a few had slipped from their bony moorings and washed into the nearby footprint of a hippo.

Sixteen vertebrae were excavated along with the boy's jaws and cranium, as were most of his ribs (some broken) and a good portion of his pelvis. It is the pelvis, with its narrow sciatic notch, that betrays the boy's gender. All of the boy's upper and lower leg bones surfaced in the excavation, as well as one upper arm bone, two forearm bones, and both S-shaped collarbones. Remarkably, one shoulder blade—a fragile element seldom well preserved in the fossil record of hominids—was found nearly intact.

The boy's teeth give evidence of his age. His wisdom teeth had not yet erupted, and he had not shed the canine ("eye") milk teeth from his upper jaw. In humans roots of adult teeth start growing before the crowns erupt and continue to grow even after the adult teeth are functioning. Most of the roots of the Turkana boy's adult teeth were between one-half and three-quarters grown. The lengths of the various roots, the emergent state of the boy's frontmost upper adult molars, and his retention of juvenile upper canines give him a likely age of about eleven years on a modern human scale of development.

The boy would have been large for a modern eleven-year-old, however, with an estimated height of about 160 centimeters (5 feet 3 inches), ignoring a slight reduction that should be assessed for the relatively flat head he shared with other early members of our genus. His large size by modern human standards suggests that he may actually have been older than eleven, and so does the maturity of his bones. Although some elements of his upper arm bone at the elbow had not turned from cartilage into bone, others had. An average modern boy attains the same degree of ossification at an age of about thirteen or thirteen and a half. In other words, by modern human standards the boy's stature and skeletal maturation were ahead of his tooth development. He may have been about eleven, as his teeth suggest, or he may have been a bit older.

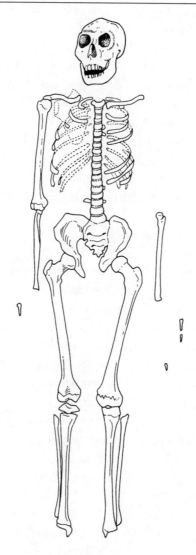

FIGURE 8.1

Skeleton of the Turkana boy, a preadolescent member of *Homo erectus*. The pelvis is narrower than that of a modern human but is similar to that of *Homo rudolfensis* (see figure 7.5).

In general physiognomy the Turkana boy resembled preadolescent males of our own species, but his visage above the neck would look strange among latter-day humans. His sloping forehead and flattened braincase would have stood out in a modern crowd, as would his heavy brow ridges. These and a low ridge along the crest of his skull served for the attachment of large facial muscles. His face was wide and short, but it sloped forward to form powerful prognathous (projecting) jaws that must have chewed with great force under the influence of the strong facial muscles. All of these features were subdued remnants of traits that were present in *Homo rudolfensis* and had been expressed even more strongly in *Australopithecus.*

The cranial features of *Homo* a million and a half years ago were actually well known before the discovery of the Turkana boy. They are displayed by three previously unearthed African skulls that range in age from about 1.8 to 1.2 million years. What the boy's bones revealed for the first time was the bodily configuration of modern human ancestors of his era.

Regardless of the boy's exact age, he was remarkably long legged (or short waisted) by modern European standards. In general proportions he was close to a modern African. My Johns Hopkins colleague Christopher Ruff has concluded that modern humans who are native to the tropics benefit from their relatively long limbs because such appendages provide a large total area of skin surface. Of all three-dimensional objects a sphere has the smallest surface area for its volume. Among modern humans Eskimos are unusually portly (relatively spherical), and they benefit from this shape by losing body heat to their environment less rapidly for a given environmental temperature than do humans who inhabit warmer climates. At the other end of the spectrum are Africans, who need to shed heat through the evaporation of sweat and therefore benefit from the long arms and legs that make them less spherical and increase their surface areas. On average their thighbones are nearly 10 percent longer than those of Caucasians, and their pelvises are nearly 10 percent narrower. In this light the Turkana boy's lengthy arms and legs make sense: he lived close to the equator.

Not only would the boy have grown to the substantial adult height of 185 centimeters (about 6 feet 1 inch), but his pelvic and leg bones were quite robust, suggesting that his lower body endured considerable stress. He presumably walked and ran great distances as a matter of everyday life. Unusually pronounced scars where muscles attached to his leg bones also point to great limb strength.

The boy's cranial capacity, estimated from an artificial cast of the inside of his skull, was about 880 cubic centimeters and would have expanded to about 909 cubic centimeters in adulthood, had he not died prematurely. This is only slightly larger than the estimate for the six- or seven-year-old KNM-ER 1590 *Homo rudolfensis* child, grown to adulthood. Estimates by Ralph Holloway for the three other known East African skulls that belong to *Homo erectus* range from 804 to 1,067 cubic centimeters in volume and yield an average that is virtually identical to the calculated capacity for the boy, grown to adulthood. These various remains range in age from about 1.8 to 1.2 million years, with the Turkana boy right in the middle.

Because the boy left so much more of his body to science than did any other member of his species, he alone reveals that from shoulder to leg *Homo erectus* was strikingly modern. Even so, his body was not identical to ours. Below the neck his skeleton differs from ours most notably in the narrowness of its pelvis. The implications of the boy's slim hips are profound. From his narrow pelvic dimensions we can show that, like a modern human, he remained physically helpless for many months after entering the world. Even if we endow his mother with a slightly broader pelvis, as would be the norm for modern humans, she would have had to give birth to him while he—and his brain—were quite small. Only a substantial postnatal interval of brain growth at the fetal rate could have allowed him to acquire his large volume of gray matter. Of course, this is hardly surprising, in that such a pattern of development had already been present in *Homo rudolfensis*.

Descent from Homo rudolfensis

Not only in its pattern of development but also in many of its skeletal features, *Homo erectus* bore a close resemblance to *Homo rudolfensis*. The pelvis of the Turkana boy, for example, is very similar to the KNM-ER 3228 pelvis that, as described in chapter 7, I used to estimate the neonatal and adult weight of *Homo rudolfensis*. In addition, the thighbones of the two species are so much alike that at one time an expert on leg bones of the human genus erroneously assigned the KNM-ER 1481a thighbone of *Homo rudolfensis* to *Homo erectus*. The similarities between the fossils of these two species range beyond bones to teeth. The cheek teeth of *Homo rudolfensis* are larger than those of *Homo erectus*, but otherwise the dental features of the two species are remarkably similar.

These many resemblances are what have convinced me that the second species evolved directly from the first. The transition from *Homo rudolfensis* must have taken place about 1.8 million years ago, because this is roughly the age of both the youngest known fossil remains of *Homo rudolfensis* and the oldest remains of *Homo erectus*. It is not clear whether evolution transformed the entire ancestral species into the descendant species or, instead, *Homo erectus* arose by way of a speciation event shortly before *Homo rudolfensis* had died out. Either way, however, the evolutionary shift here was minor compared to the earlier divergence of *Homo* from the much more apelike *Australopithecus*.

The Dehumanizing of "Homo" habilis

Much controversy has surrounded another member of the human family that shared Africa with *Homo rudolfensis* from about 2 to 1.8 million years ago and then with *Homo erectus* from about 1.8 to 1.6 million years ago. This was *"Homo" habilis*. I place the name of the genus in quotation marks because I join many other workers in excluding this creature from the human genus, where it was placed at the time of its scientific christening. Although this species now seems clearly to represent an evolutionary deadend, off the main line of human evolution, its place in the annals of anthropology is so prominent that I must digress from my narrative path toward modern humans to review its history.

The name *Homo habilis* is officially attached to a portion of a cranium and most of a lower jaw that Mary and Louis Leakey's son Jonathan discovered in 1960 at Olduvai Gorge in strata now known to be about 1.8 million years old. The stage of tooth development in the lower jaw of this creature matches that of a modern human eleven or twelve years old. The next few years brought additional fossil finds at Olduvai that appeared to represent the same species. Louis Leakey, Phillip Tobias, and John Napier chose the label *"Homo" habilis*, meaning "handyman," for this species because these early workers assumed it to have been the maker of what were then the oldest known artifacts in the world, the stone tools of the Oldowan culture that we now attribute to *Homo rudolfensis*. The first remains of *"Homo" habilis* to be described were thought to predate *Homo erectus*, and by the late 1960s, the era of the single-species hypothesis, the human family tree looked quite simple and unbranched. It seemed that *Australopithecus* had turned into *"Homo" habilis* and that this first member of our genus had then evolved into *Homo erectus*, which, in turn, became *Homo sapiens*. As so often happens in science, however, new facts challenged a simple view of nature.

In the case of *"Homo" habilis,* the unsettling new data came from skulls. From Olduvai and the vicinity of Lake Turkana over the years there have come crania that, nearly everyone agrees, belong to *"Homo" habilis,* although their capacities are quite small. Estimates for three of these—OH 24, OH 13, and KNM-ER 1805— are 450, 500, and 600 cubic centimeters, respectively. This range of values overlaps substantially with the range for *Australopithecus.* Beginning in the 1970s, discoveries of other skulls to the north of Olduvai, along the shores of Lake Turkana, began to convince many researchers that the human family tree actually lived up to its name: it was branched. What turned up were skulls with cranial cavities much larger than estimates for *"Homo" habilis* from Olduvai. These big-brained forms included the KNM-ER 1470 and 1590 skulls, which I described in chapter 7 as belonging to *Homo rudolfensis.* So large were the caps of these skulls that some workers immediately accorded them status as a distinct species, but others took a more conservative stand consistent with the single-species hypothesis. In their view what we now recognize as *Homo rudolfensis* was simply the male form of *"Homo" habilis,* with a much larger head than his mate—a head that matched his much larger body. A lengthy controversy ensued: Did *"Homo" habilis* share its world with a separate species of *Homo?*

The debate took on a new dimension with the unearthing of a relatively complete postcranial skeleton of *"Homo" habilis* from deposits about 1.6 million years old. This was the OH 62 skeleton from Olduvai that Donald Johanson and his colleagues described in 1987 and that Johanson celebrated in the book *Lucy's Child,* coauthored with James Shreeve.

The OH 62 skeleton of *"Homo" habilis* is remarkable not only for its completeness—parts of the skull, right arm, and both legs are preserved—but also for its proportions. Although living nearly a million and a half years after Lucy, the OH 62 animal was more apelike in form than she. Among primates the lengths of arms relative to legs vary with body size. (We can easily observe that tall people tend to be long legged even for their height, whereas short people tend to be long waisted.) To compare the proportions for different species we must correct for disparities of individual body size. One way to do this is to examine animals that are roughly the same size. For example, we can compare a pygmy chimpanzee, a human Pygmy, Lucy, and OH 62. For an average pygmy chimp, the humerus (upper arm bone) is about 97 percent as long as the femur (thighbone), whereas for an average modern Pygmy the ratio is only about 74 percent. For Lucy it is just about halfway in between — 85 percent. On the other hand, the ratio for OH 62, though not precisely known because her preserved limb bones are incomplete, appears to have been nearly as large as for a pygmy chimp — about 95 percent! In other words, OH 62 had gangling arms and stubby legs. This younger animal, with an estimated weight of 25 kilograms (55 pounds), was about 20 percent lighter than Lucy. Almost certainly, the diminutive OH 62 was a female, and it stands to reason that she put her long arms to good use in climbing trees. I will henceforth refer to her species as *"Homo" habilis*, because I view it as unqualified for status within our genus.

In interpreting the OH 62 animal, Johanson and his colleagues clung to the view that she was the female counterpart of the big-brained creatures now known as *Homo rudolfensis*. This appraisal was based on dental similarities, but other workers saw the difference in brain size as a problem: How could the females of a species be such pea brains compared to the males? At least as problematical is the striking contrast between the tiny, apelike physiognomy of OH 62 and the large, markedly human-looking pelvis and thighbones that are now assigned to *Homo rudolfensis*. With *Homo rudolfensis* being nearly two and a half times as heavy as OH 62, even the estimated body weights for these forms are too disparate to wedge them into a single species. In 1989 a report of a new fossil find dealt a fatal blow to this facet of the single-species hypothesis. This was the discovery by Richard Leakey and his colleagues of a fragmentary skeleton, labeled KNM-ER 3735, east of Lake

Turkana. It represents a large animal—one that weighed an estimated 46 kilograms (about 100 pounds), compared to about 25 kilograms for OH 62—yet it possessed similar, apelike limb proportions—relatively long, powerful arms and short legs. Not surprisingly, the few preserved skull fragments of KNM-ER 3735 also resemble those of the relatively small-brained *"Homo" habilis*. Here we apparently have a male of this species, one strikingly more apelike than the markedly human *Homo rudolfensis*.

Out of Africa

Neither *"Homo" habilis* nor *Homo rudolfensis* appears ever to have ventured from the African continent. *Homo erectus* was the first member of the human family to make its way northward from Africa, and it did so very early in its history. The traditional view was that the diaspora took place about a million years ago, but two new developments have overturned this chronology. One is the discovery, between the Caspian Sea and the Black Sea, of a jaw of *Homo erectus* that dates to between 1.6 and 1.8 million years ago. The other is the calculation of a new radiometric age of about 1.8 million years for remains of *Homo erectus* that were collected long ago in Java. It seems that this species, the first hominid to escape from Africa, did so very soon after making its appearance and then, skirting western Europe, traveled far to the east. In fact, *Homo erectus* may never have occupied western Europe. All that we know for sure is that some of its populations wended their way eastward, through Southeast Asia to what is now Indonesia. The magnificent fossils they left behind in China were at first dubbed "Peking man," and their Indonesian remains became known as "Java man." In 1894 Eugene Dubois established the specific name *erectus* for the Javan fossils, but he saw the species as being so different from modern humans that he originally placed it in the separate genus *Pithecanthropus*, meaning "ape human."

Dubois had abandoned a promising academic career in Holland at the age of twenty-nine to search for human ancestors in the Dutch East Indies, where he assumed a convenient post as a medical officer in the army. He agreed with Charles Darwin that the crucible of human evolution was likely to have been a region of the world apes inhabit today. Darwin's first choice was Africa, but Dubois favored Southeast Asia and Indonesia, partly because this region was home not only to the orangutan but also to the gibbon, which he believed to be a close human relation. The Dutch presence in Indonesia undoubtedly influenced Dubois's choice of a field area, as did the fact that this region had yielded fossil mammals which he believed might have been contemporaries of human ancestors.

Although *Homo* did not actually originate in Indonesia, Dubois's reasoning bore fruit. Excavations that he oversaw on Java yielded the first discoveries of what we now call *Homo erectus*. These included a skullcap with the now familiar heavy brow and low, sloping forehead. To a scientific world that at the time was aware of no hominid fossils except for ones representing Neanderthals, the primitive skull from Java was an extraordinary find. Ralph Holloway's modern estimate for the cranial capacity of the Javan skull is 940 cubic centimeters, and Dubois's calculation made it only about 5 percent larger. Dubois's skull and others that have since surfaced in Java are clearly less than a million years old, but how much less we still do not know.

From farther north, in China, has come an array of *Homo erectus* specimens that date back only a few hundred thousand years. The most important of these by far were collected from cave deposits at Zhoukoudian, near Beijing, in the 1920s and 1930s. There is no way of dating the Zhoukoudian remains precisely, but they range in age from about 500,000 to 250,000 years. The German anthropologist G. H. R. von Koenigswald, who had reported on discoveries of *Homo erectus* in Java, and Franz Weidenreich, a German anatomist who was briefly affiliated with the University of

Chicago, agreed that the specimens from Java and China closely resembled each other. At the time the Javan fossils still bore the label *Pithecanthropus erectus*, and the Chinese fossils, as a result of provincial attitudes, remained segregated not only as a separate species but also as a separate genus; they were widely known as *Sinanthropus pekinesis* ("Chinese human from Peking").

The Chinese fossil vanished with the onset of World War II. Weidenreich, who had been studying the fossils in China, feared hostilities from the Japanese occupation forces and decided to relocate to New York in 1941. The Chinese sought similar refuge for their fossils and placed them in the hands of United States Marines. The fossils arrived in Beijing on December 7, the very day of the attack on Pearl Harbor. The Japanese quickly captured the Marine custodians, and the fossils were never seen again. Fortunately, a total tragedy was averted. With foresight that proved invaluable to anthropology, Weidenreich had seen fit to fabricate excellent replicas of the fossils, which still reside in New York's American Museum of Natural History; the shapes of the Zhoukoudian fossils were not lost to science.

Evolutionary Stagnation

Homo erectus ranged far during its long history, but it did not evolve very much. So distant are Java and China from central Africa that it is easy to forget that the Far Eastern fossils have been assigned to the same species as the Turkana boy, who lived more than a thousand miles to the west and much earlier in time. The Chinese fossils are more than a million years younger than the boy and about a million and a half years younger than the oldest remains of *Homo erectus* from Java. Taken together, these various remains document a period of relative evolutionary stability in our ancestry. Some workers have argued otherwise, suggesting that the brain of *Homo erectus* expanded significantly during the lengthy history of the species. These workers have assumed that a few large skullcaps from young deposits in Java belong to *Homo erectus*. As I will discuss later, I view them as early members of another, brainier species, *Homo neanderthalensis*. If we exclude these fossils from *Homo erectus*, then no significant brain expansion is evident for this species during its long stay on earth. An African skull of *Homo erectus* from Olduvai, labeled OH 9, is well dated as far back as 1.25 million years, yet it exhibits quite a large cranial capacity, estimated at 1,067 cubic centimeters. This is slightly above the average for the Zhoukoudian crania, which are somewhere between three-quarters of a million years and a million years younger.

The stone tools of *Homo erectus* were as distinctive as its bones. About a million and a half years ago, the more complex stone culture of this species replaced the Oldowan. This was the Acheulean industry of *Homo erectus*. The simple Oldowan industry had failed to progress in any apparent way: the youngest artifacts closely resemble the oldest. Forty or fifty thousand successive generations of primates came and went in a society that experienced general evolutionary stagnation. *Homo erectus* moved a giant step ahead by generating the Acheulean culture, but not when it first appeared on the scene. The species' oldest known tools have come from strata about 1.4 million years old, a few hundred thousand years younger than the oldest known bones of the species. *Homo erectus* apparently failed to achieve a technological breakthrough immediately after it came into being.

The strata that have yielded remains of *Homo erectus* in China and Java appear to contain no Acheulean tools like those found far to the west. The apparent explanation is simply that the population that migrated from Africa left their continent before the stone culture was invented and then failed to generate a comparable industry. Support for this chronology comes from the discovery of stone tools of a primitive Oldowan type with the jaw of *Homo erectus*, found between the Caspian and Black seas and dating to between 1.6 and 1.8 million years ago.

The implements of the western populations of *Homo erectus* are labeled Acheulean. The most characteristic of these artifacts are palm-sized, flattish cleavers and hand axes that are bifacial, meaning that their edges were formed by chipping on two sides. The typical cleaver terminated in a straight cutting edge, like a modern hatchet. It was not mounted on a handle, however, but simply held in the hand. The hand axes, as their name implies, were also handheld but were more pointed.

Although the Acheulean tools of *Homo erectus* were more complex and generally larger than the Oldowan flakes of *Homo rudolfensis*, they were still relatively simple. In addition, *Homo erectus* made little use of bone, which it might have fashioned into spear points, fishhooks, awls, or more delicate implements. Furthermore, even the stone tools, like those of the Oldowan culture, failed to advance appreciably in form. In other words, the Acheulean culture, like the creature that created it, experienced evolutionary stability for more than a million years.

One might hope that the bones and artifacts of *Homo erectus* would indicate more about its mode of life than those of early *Homo* reveal, but here they disappoint us. Marks of the stone tools of *Homo erectus* are evident on a few fossil bones of species that were possibly its prey, but whether our ancestor was the killer of large animals or the scavenger of their carcasses remains unclear. In any event, *Homo erectus* seems not to have attached stone points to projectiles.

There is better evidence of a cultural advance for *Homo erectus* in the use of fire. Fire in the open can result from lightning, but humans generally account for its presence in caves. Apparent hearths in caves date back only about half a million years, to late in the time of *Homo erectus*, but we have evidence that hominids harnessed fire nearly a million years earlier. At Lake Baringo, in Kenya, stone tools of this vintage occur with lumps of burned clay and with stones arranged in the configuration of an open hearth. It appears that, at least late in its history, *Homo erectus* roasted its meat.

Neanderthals: Another Step

From the shores of the Solo River in Java have come problematical fossil crania that, although poorly dated, are obviously younger than the ones belonging unequivocally to *Homo erectus*. The Solo crania had somewhat larger cranial capacities than the average for *Homo erectus*. To some other workers they appear to represent a younger species that includes the famous Neanderthal race of Europe and the so-called Rhodesian man of Africa; other workers refer to them as archaic *Homo sapiens*. In my view they deserve status as a distinct species, for which the name *Homo neanderthalensis* apparently has precedence. In this form, we see the results of another evolutionary step in the history of *Homo*.

A new radiometric date for the so-called Bodo skull from Ethiopia appears to establish the presence of *Homo neanderthalensis* in northern Africa about 600,000 years ago. Actually, the preservation of the Bodo skull is too poor for experts to rule out the possibility that it represents a species distinct from and older than *Homo neanderthalensis*. The earliest known hominid fossils from Europe are also of uncertain status. This is a large group of teeth and fragments of jaws and braincases that come from cave deposits of the Atapuerca region of northern Spain more than 780,000 years old. These remains display a number of primitive features and may represent direct ancestors of the so-called classic Neanderthals. Adding to the confusion surrounding this early population is its stone tool industry, which generated only simple flakes that, surprisingly, are more primitive than much older Acheulean implements of *Homo erectus*.

There is no question that the Neanderthal lineage was established in Europe slightly before 300,000 years ago. This is the age of three remarkable skulls that in 1993 a group of Spanish anthropologists led by Juan-Louis Arsuaga described from another cave in the Atapuerca region. These skulls display a large range of cranial capacities, the smallest being slightly above 1,100 cubic centimeters and the largest approaching 1,400. (Recall that the average for modern humans is about 1,320.) These fossils are the earliest remains of Homo neanderthalensis yet found in Europe. Even so, they share many features with the classic Neanderthals of Europe, whose earliest remains are only about 230,000 years old. The Atapuerca skulls also resemble less well preserved ones from Africa ("Rhodesian man") and from Europe ("Heidelberg man" and others) that also antedate the classic Neanderthals. The skulls from the banks of the Solo River are obviously closely related to these. All of these fossils may represent Homo neanderthalensis.

Homo erectus, the parent species of Homo neanderthalensis, persisted even after the descendant species was established. While the archaic Neanderthals occupied Spain, the smaller-brained Zhoukoudian population of Homo erectus lived on in China until somewhere between 250,000 and 150,000 years ago.

Who, then, was the big-brained Homo neanderthalensis? We can answer this question in some detail for the classic Neanderthals of Europe because these creatures left a magnificent fossil record, apparently by design: they buried their dead. Powerful evidence of this behavior is the fetal position of many of the preserved corpses. The frequent burial of Neanderthals with animal bones, probably carrying meat, may even indicate belief in an afterlife that required provisions.

The classic Neanderthals ranged from England and Spain, in the west, nearly to China, in the east. They lived from about 230,000 to 30,000 years ago. Their brains were actually larger than ours, with an average cranial capacity slightly above 1,500 cubic centimeters (modern humans average about 1,320). We need not feel intimidated. Modern humans who inhabited Europe about forty thousand years ago had brains slightly larger even than the Neanderthals. Like the Neanderthals, they seem also to have had very heavy bodies. Their larger brains may have been devoted to operating heavier muscles than ours, not to entertaining heavier thoughts.

Neanderthals' crania were larger than ours, but they were not shaped in the same way. Behind the massive brow ridges, a long, sloping forehead formed the front of a remarkably flat-topped braincase. It appears that less of the Neanderthal brain than ours was concentrated in the frontal lobes, which may imply lesser intelligence. Much debate has centered on whether Neanderthals had the capacity to speak. The configuration of their skull seems to indicate a forward-projecting posture for the neck, within which the larynx had not descended as far as ours, but their level of verbal articulation remains in doubt.

The need to devote a large volume of brain tissue to motor activities is evident in the robustness of the Neanderthal skeleton. An average male stood only about 169 centimeters tall (5 1/2 feet), but he was quite sturdy, probably outweighing a modern human having a comparable amount of body fat. He was long waisted, with a large body and short legs for his height in comparison to a modern man. Deep scars on Neanderthal bones betray the attachment of heavy muscles. Neanderthals were much brawnier, on average, than are modern humans.

By our standards Neanderthals also displayed a miserably weak chin beneath a strongly projecting mouth. In these features, and in the general shape of the cranium, they were closer in form to *Homo erectus* than to us, with our tall forehead, flat face, and sharp chin. Badly worn front teeth of many Neanderthals point to a habit of chewing on something other than food—probably leather that required softening for one purpose or another. A usual pattern of heavier wear on the right side of the tooth row indicates a tendency toward right-handedness.

Neanderthals shared with us a capacious pelvis. By comparison the more ancient Turkana boy was remarkably slim hipped. The Neanderthals' wide pelvis, like ours, obviously was related to the presence of a big brain. The pelvis permitted babies to be born at a large size and, hence, to have big heads and brains. For Neanderthals, as for us, a wide pelvis lengthened the interval of brain

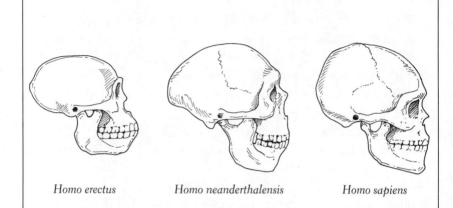

Homo erectus   Homo neanderthalensis   Homo sapiens

FIGURE 8.2
Skulls of three species of *Homo*. In many ways the skull of *Homo neanderthalensis* more closely resembles that of *Homo erectus* (from Java) than that of *Homo sapiens*, with its tall forehead, flat face, and sharp chin.

growth within the womb. The wide pelvis allowed for brain expansion without prolongation of the interval of physical helplessness after birth much beyond its duration in earlier species of *Homo*. It is easy to see why in Neanderthals, as in modern humans, females had broader hips for their size than males. The pelvic broadening of Neanderthals and humans was not achieved without sacrifice. The narrow pelvis of the Turkana boy gave his hip muscles a mechanical advantage over ours in supporting his body as he walked and ran.

During most of their existence in Europe, the classic Neanderthals lived in the vicinity of continental glaciers. The projecting nasal region of their skull suggests the presence of a very large nose, which may have served well for warming and moistening the frigid, dry air of Ice Age winters.

Neanderthals surpassed *Homo erectus* in the manufacture of

tools. Among their stone productions were crude knives, single-edged scrapers, and triangular and heart-shaped hand axes. Only a small percentage of their implements show wear from attachment to a handle. The Neanderthal stone culture, known as Mousterian, is widespread throughout Europe, southwestern Asia, and Africa, but, like earlier stone cultures, it displays little evidence of technological change through time. Neanderthals seem to have been culturally stagnant, at least when it came to fabricating tools. They rarely used bone as a raw material. In addition, the bedrock from which they fashioned their tools invariably lay close to the final resting place of the tools, indicating that the manufacturers engaged in little migration or trade.

In fact, Neanderthals endured a harsh life close by the frigid glaciers of Europe. Study of their skeletons reveals that few individuals reached the modest age of forty years, and many endured severe injuries and degenerative diseases during their short lives. Despite their travails, these curious close relatives of ours displayed compassion for their disabled and elderly, who by our standards were barely middle-aged. Pathological studies of nine Neanderthal skeletons from Shanidar Cave in Iraq have revealed that one male had apparently received a blow to the left side of his head that fractured his eye socket and probably blinded him. Another blow had so damaged the right side of his body that his collarbone, shoulder blade, and upper arm subsequently withered. His right hand was also lost and his lower right leg, severely damaged. This crippled individual survived for several years after his terrible mishap, obviously through the charity of others.

Neanderthals may have had empathy, but they lacked art. They created very few art objects per se and virtually never adorned their tools with symbolic or decorative patterns. This creative deficiency has contributed to the Neanderthals' reputation as intellectual dwarfs, although we cannot assess the accuracy of this appraisal.

Finally, Homo sapiens

However our species may stack up against the Neanderthals intellectually, we were present long before they died out. Here again we have evidence of a temporal overlap between two species of *Homo*. Exactly when *Homo sapiens* came into existence we cannot say, but fossils, artifacts, and genetic data all point to the origin of the species in Africa more than 100,000 years ago, long before the Neanderthals had disappeared from Europe. Several African sites have yielded early fossil skulls of *Homo sapiens*, and a small number of the fossils are between 100,000 and 120,000 years old. These skulls, some of which are fragmentary, clearly exhibit our high-vaulted forehead, sunken cheeks, and sharp chin. Quite recently, several sites in Zaire have yielded a magnificent array of bone artifacts, including barbed and unbarbed spear points, showing that by at least 90,000 years ago a species which must have been ours was attaching elegant spear points to shafts; its culture had advanced far beyond that of the Neanderthals. So meager is the collection of fossil remains of early *Homo sapiens* in Africa that our species may conceivably have an undetected history extending back 200,000 years or more from the present.

In recent years controversy has surrounded ostensible genetic evidence that all modern human populations trace back to a single African female, aptly christened Eve. Ancestry as narrow as this for modern *Homo sapiens* lacks convincing support, but genetic evidence for an origin somewhere in Africa remains strong. The most basic evidence comes from genetic variation, as indicated by blood proteins and enzymes. These are genetically coded compounds that diversified through time within all human populations after the populations came into existence. The greatest genetic divergence has occurred in Africa, indicating that populations there have the longest history. The European and Asian races exhibit less genetic diversity and are more closely related to each other than to Africans, apparently having had relatively little time to diverge genetically from one another.

What we do know about the early history of *Homo sapiens* is that by some 100,000 years ago it had traversed the narrow neck of land between the Nile delta and the Sinai Peninsula, along which, more than a million and a half years earlier, *Homo erectus* had wended its way northward. From this crossing our species eventually spread throughout the non-African world, but it seems first to have remained for a time in the Middle East.

Overlap with the Neanderthals

The oldest Middle Eastern fossils that represent our species were found during the 1930s in what is now Israel. Some are from Skhül, a cave in Mount Carmel; others are from Qafzeh, a second cave to the south of the town of Nazareth. Some of the preserved skulls display brow ridges that are stronger than those of all but a few modern humans but weaker and differently shaped than those of Neanderthals. Others of the skulls lack these archaic features. Application of new dating techniques to the cave deposits of Skhül and Qafzeh in the late 1980s jolted the anthropological community. The skulls turned out to be unexpectedly ancient, some 80,000 to 100,000 years old. The new dates had profound evolutionary implications because Neanderthal skeletons from two other Israeli caves, one of them also in Mount Carmel, turned out to be much younger, between 40,000 and 60,000 years old. Other Neanderthals had populated the region long before, between about 200,000 and 90,000 years ago. If the dating of the various fossils has been accurate, then Neanderthals and *Homo sapiens* coexisted for a time in Israel. Perhaps the two species lived in harmony. In any event, the evidence deflates the multiregional model—the idea that Neanderthals and Neanderthal-like populations evolved gradually into modern humans throughout the broad region comprised of Africa, Europe, and Asia. Neanderthals obviously did not disappear by turning into *Homo sapiens,* but rather shared the world with us for tens of thousands of years. In fact, modern humans may well have had a separate origin altogether from the Neanderthals; we may have evolved from a different population of *Homo erectus* than the one that turned into them.

The range of sexual behavior among modern humans would suggest that Neanderthals and *Homo sapiens* probably engaged in a bit of interbreeding. It is nonetheless difficult to imagine that the two generally found each other sexually appealing. One might also doubt whether any hybrid offspring, if biologically viable, would have been welcomed into the society of either species. The fossil record in Europe also suggests that the two hominid species were socially incompatible, at least under some circumstances.

The European history of *Homo sapiens* is entirely within the range of radiocarbon dating. It extends back only about 33,000 or 34,000 years, to the time when our species seems originally to have migrated northward from the Middle East. Neanderthals were already well established in Europe, but within just three or four thousand years after the arrival of *Homo sapiens* they vanished. No Neanderthals date to less than 30,000 years ago. The idea that their disappearance had nothing to do with the arrival of *Homo sapiens* strains credulity. After all, the newly discovered skulls from Atapuerca, Spain, show that the Neanderthal lineage had occupied Europe for more than a quarter of a million years before its sudden demise. To put it another way, having survived for more than ten thousand generations, the Neanderthals disappeared in the course of just one or two hundred generations once *Homo sapiens* were on the scene.

It might be tempting to conclude that our species intentionally killed off Neanderthal, either as its conqueror in war who took no prisoners or as its hunter. This, however, would be leaping to a conclusion. An alternative possibility is that *Homo sapiens* displaced Neanderthals in competition for food, hunting grounds, or cave shelters. The Neanderthals disappeared when the world had already plunged far into the most recent interval of glacial expansion during the modern Ice Age. They had endured similar episodes in the past, but not in the presence of our species. We must entertain the possibility that it was by some combination of aggression and competition for resources that we did in the Neanderthals.

Cro-Magnons and Creativity

In the world of anthropology, these early *Homo sapiens* in Europe go by the name of Cro-Magnon. The Cro-Magnon people were larger than modern Europeans, with men averaging about 183 centimeters (6 feet) in height, but otherwise the Cro-Magnons differed from their modern descendants in only very minor skeletal features. Their culture quickly soared far above that of the Neanderthals. Cro-Magnons obviously brought advanced technology with them from Africa. They fashioned artifacts not only from stone but also from bone, antler, and the ivory of mammoth tusks. They transported raw materials, including seashells, vast distances. Elegant stone projectile points reveal that they were sophisticated big-game hunters.

Early members of our species invented art as we know it, rather than devoting virtually all of their constructive activities to tools and other necessities, as Neanderthals were compelled, or inclined, to do. The Cro-Magnons' radiant cave paintings of the animals that shared their world have come to epitomize for us their creative spirit. The Cro-Magnon people seem also to have developed a hierarchical social structure, burying a few of their revered citizens in elaborately beaded costumes, and they painted and sculpted symbolic figures that obviously served religious functions.

What set early *Homo sapiens* apart from the millions of species that had preceded it on earth was that it engineered a constantly changing relationship to its environment: it created an evolving culture. Further, the culture was a patchwork quilt. Populations in different regions modified their local cultures in unique ways and evidently traded regional goods and materials to their mutual benefit. Yet, even with such diversity, there were pervasive trends across much of Europe.

Anthropologists recognize distinct stages in the evolving Cro-Magnon culture for the brief period between 34,000 and 10,000 years ago, which was only about 10 percent as long as the Neanderthals' entire stay on earth. Small animal carvings are among the earliest Cro-Magnon artifacts. "Venus" figurines, presumed to be fertility totems, came later, and the magnificent cave paintings that are so frequently reproduced are younger than about 18,000 years. In parallel with their art, the tools of Cro-Magnon evolved greater sophistication. Their predecessors, the Neanderthals, like earlier species of *Homo*, had experienced little technological progress. Neanderthals were locked in a struggle with their environment that typically ended their lives early, and they failed to mold this struggle to their benefit by generating a highly dynamic culture. The unique achievement of early *Homo sapiens* was to create such a culture—to exert progressive control over its environment.

Today of course we debate the degree to which we, the immediate descendants of Cro-Magnon, should rein in our powerful creativity. Two and a half million years ago, early *Homo* took the first step toward our boundless creativity by napping away at shapeless stones to gain a sharper edge for performing some menial task in its newfound life on the ground. Little did it know that it could grasp the stones it shaped only because its distant ancestors had evolved prehensile forelimbs that served to clutch branches high in the trees. Little did it know that it was inaugurating a culture on the ground that someday would threaten to spin out of control.

CHAPTER NINE

An Unlikely Birth, a Dubious Future

We cannot know ourselves deep to the core without understanding our origins. When we dig into our evolutionary genealogy, our motives resemble those of an adopted child who, in the course of an otherwise happy life, seeks to identify biological ancestors. In our evolutionary search we learn that some of our peculiarities are a matter of heritage. Natural selection produced certain traits to the benefit of our distant ancestors and then failed to remove these features even after they had lost their value. On balance our wisdom teeth, for example, probably do more harm than good in a modern jaw that, more often than not, proves too small to accommodate them. Other human traits have been flawed or problematical from the very start—products of evolutionary compromise or of some other deficiency of natural selection. A more perfect process would have done better than to saddle us with infants who cannot walk unattended for their first fifteen months of life. Maturation as rapid as that of other animals would be much to our advantage.

A somewhat thornier question is whether natural selection has embedded in our genes certain patterns of behavior that we seem unable to shake even when we apply what we would like to believe is free will. Just how aggressive are we deep down inside?

How fundamentally competitive? How willing to make sacrifices for the benefit of others? To what extent are our predilections in all of these areas a matter of genetics?

We seem endowed with innate curiosity as to how our forebears guided culture to its present state. Those who have gone before us invented creation myths to justify their lifestyles and, in fact, their very existence. Numerous myths outside the Judeo-Christian world offer alternatives to Genesis, which itself presents more than one creation story. As we steer civilization into the future, we gain perspective by looking over our shoulder to the path already taken—the actual path of our biological and cultural evolution—as opposed to the fanciful paths portrayed in creation myths. Our perspective broadens when we view the journey in context, assessing the environmental changes that our ancestors either endured or engineered along the way. Sometimes nature propelled them in a certain direction, and sometimes they cleared a route for themselves, controlling their own fate.

In this book our central story has been about the sudden origin of the human genus from a more apelike creature with a much smaller brain. Five primary themes—none of them part of the conventional wisdom of anthropology—have emerged, all of them with implications for the way in which we view our position in the world.

First, the large brain of early Homo *was not the product of a long, gradual evolutionary trend but of a geologically abrupt transition.* This conclusion runs counter to the traditional idea that persistent natural selection expanded the brain, almost inexorably, over hundreds of thousands of years. On the contrary, the rapid encephalization forms part of a body of new evidence that makes the pattern of human evolution look highly punctuational.

Second, the big brain of Homo *came into being not because such an organ became useful for the first time but because its evolution became possible for the first time.* What happened was that a change of vegetation in Africa removed the constraint that had held *Australopithecus* in a state of evolutionary stagnation for more than a million and a half years.

Third, our genus was born of an environmental crisis, which means that it might never have been born at all. The shrinkage of the woodland habitats that had sheltered *Australopithecus* for more than a million and a half years might well have driven every population of the ancestral genus to extinction. Luck must have played a role in the survival of one population long enough to experience the genetic changes that turned it into *Homo*.

Fourth, the origin of Homo *is traceable to a local geological event that had global consequences.* Remarkably, the evolution of the human genus resulted from a chain of events that began with movements of the seafloor an ocean away from Africa—movements resulting from forces generated deep within the earth.

Fifth, the animal with the big brain that emerged from the Ice Age crisis was advanced but also flawed—the imperfect product of an imperfect process. This process, natural selection, negotiated a profound compromise in creating the human genus. It compelled members of the genus to cope with helpless infants because growth of the big brain of *Homo* was contingent upon this kind of sacrifice. The value of the brain for survival and reproduction obviously outweighed the cost of tending the problematic offspring, but the cost remained.

If our biological imperfection is humbling, the accidental nature of our evolutionary birth is astounding. Had a skinny dam of land not happened to rise from the depths to separate the Atlantic Ocean from the Pacific, then the chain of events that triggered the evolution of *Homo* would never have begun.

The element of chance in our appearance on the earth bears on the way in which we view our place in nature. Ignoring the evident biblical concept of stewardship, some modern humans claim Old Testament authorization for human dominion over nature. Implicit here is the notion that, as part of a grand plan, we humans were somehow placed on the earth with a special purpose that included conquest of the natural world. Even humans of less fundamentalist bent tend to see themselves as essential components of a complete planet Earth. This view is in part a remnant of the pre-Darwinian concept of a Great Chain of Being—a perfectly created ladder of life in which humans stood proudly on a rung above all other earthly beings, closest to the creator. Actually, we are not essential to planet Earth, our Ice Age birth proclaims, but newly arrived interlopers.

Interlopers we are indeed. Ironically, we are now engendering environmental changes that threaten to alter the very climatic regime whose origin brought our genus into existence. The fortuitous appearance of *Homo* deserves closer scrutiny, as does the trouble-laden proclivity of modern humans to modify our environment—and ourselves.

Terrestrial Imperative/Catastrophic Birth

The evolutionary stagnation of *Australopithecus* through some hundred thousand generations was the first of the three mysteries that I introduced at the beginning of this book. *Australopithecus*, with a brain only slightly larger than that of a chimpanzee, would have benefited by evolving the much larger brain of *Homo* at any time during its long stay on earth. The problem for *Australopithecus* was that the evolution of such a brain would automatically have produced immature offspring unable to cling to their semiarboreal mothers, and no mother whose arms were occupied with climbing could have clutched such an infant. Here, I have emphasized, is an explanation for the remarkable failure of *Australopithecus* to evolve the large brain of *Homo* for more than a million and a half years—a failure that has previously gone unexplained. In effect, a kind of catch-22 was operating. *Australopithecus* needed to be fully on the ground in order to evolve a large,

Homo-like brain, and yet it would have needed such a brain to live a successful life fully on the ground.

It took an environmental crisis to remove the comfortable straitjacket—to force *Australopithecus* to abandon its habitual climbing and conform to the terrestrial imperative for the evolution of the big brain of *Homo*. This crisis issued from climatic change at the onset of the Ice Age. A widespread drying of Africa shrank and fragmented the woodland habitat that had for so long protected *Australopithecus*. In effect, nature threw *Australopithecus* to the lions.

I have emphasized that the shift to a totally terrestrial way of life, though fraught with danger, was a simple locomotory step for a population of *Australopithecus* because members of the genus were already part-time ground dwellers. In other words, these animals were physically capable of moving about on the ground, even if they suffered predation from faster running carnivores, against which they initially had little defense when there were no large stands of trees for refuge. At some point during the crisis that the Ice Age brought on, a group of australopithecines must have abandoned its natural inclination to climb trees every day. Its members simply resigned themselves to the harsh new environmental regime and concentrated on making the most of it. The result— by no means an inevitable one—was the catastrophic origin of *Homo*.

There were two steps in the transition. First came the involuntary change of behavior that prevented *Australopithecus* from continuing to rely on trees for refuge. Then came the evolutionary transformation of one population of *Homo*. This sequence of events is a prime example of the generalization that in the evolution of higher animals a change of behavior tends to precede a major change of physiognomy. This is the evolutionary equivalent of form following function.

Although confinement to the ground must at first have been

compulsory when refugial forests became scarce, in time natural selection purged early *Homo* of genetically based urges to climb, just as it eventually altered the animal's bodily features in improving its ability to cope with the new circumstances.

How a Dying Species Gave Birth

In the environmental crisis that the Ice Age brought on in Africa, we have found answers to both of the two additional mysteries I posed in chapter 1. The second was the geologically sudden decline to extinction of *Australopithecus*, and the third was the similarly abrupt evolution of one of the late-surviving populations of *Australopithecus* into the genus *Homo*.

The death of a species is seldom sudden and painless. When intensified predation plays a major role, for example, local populations of the victim species variously fragment and shrink. Then, one after another, they die out. The process may drag out over thousands, tens of thousands, or even hundreds of thousands of years, and the last survivors of any one population may actually succumb not to predators but to disease or drought or any of a spate of other normal agents of mortality. Predation will simply have tipped the balance in a fatal direction, weighing in with the many other destructive agents against which the species had previously managed to endure.

In assessing the demographic impact of the climatic changes that I believe drove *Australopithecus* out of the trees, I have emphasized the problems that were imposed by the reproductive pattern of so large a primate. As chimpanzees and humans illustrate, large primates take a long time to reach reproductive maturity, and they give birth to few offspring before they die. In effect, females put all of their eggs in very few baskets, if not just one, and they expend a great deal of time and energy in rearing their precious progeny. For animals of this kind—and certainly *Australopithecus* was one—very high rates of infant and juvenile mortality are intolerable. The heavy toll that carnivores took as the protective forests shrank early in the Ice Age must have sent many a population of *Australopithecus* into a downward spiral. The delicate balance that weighed births against deaths would quickly have tilted toward extinction.

Even if *Australopithecus* suffered a population crash at the onset of the Ice Age, without further evidence we cannot rule out the possibility that its entire constricted population turned into *Homo*. "Bottlenecking" is the word I use for this kind of species-to-species transition. The alternative is that just one population of the beleaguered *Australopithecus* evolved into *Homo*, while others survived for a time alongside the new genus, though perhaps not in the same region; this would have amounted to evolutionary branching, or speciation. Unfortunately, the known fossil record cannot at present resolve the issue of overlap.

Whether *Homo* evolved through speciation or bottlenecking, those populations that failed to take part in the transition died out. When the last of the populations not transformed into *Homo* were gone, so was *Australopithecus*. Regardless of the particulars of the emergence of *Homo*, there had to be a measure of luck in the survival of one population long enough to make the evolutionary transition.

Giving Carnivores Their Due

Jack Stern and Randall Susman deserve most of the credit for formulating the argument that the defenseless australopithecines lived a semiarboreal life. In many significant aspects these animals' upper limbs were more apelike than human, and their long, curved toes would have served better than ours for climbing trees but less well for running on the ground. Various skeletal features of *Australopithecus*, from its inner ears to its toes, show that it lacked our abilities on the ground—our speed, endurance, and balance. It is important to understand, however, that while the shapes of bones are quite revealing, additional evidence comes from the dangers that we know *Australopithecus* would have faced on the ground.

Anyone who would argue that *Australopithecus* made no use of its evident climbing skills is ignoring the ecological facts of life in Africa. For forest-dwelling antelopes the most damaging result of the early Ice Age was probably a shortage of food resources, but for *Australopithecus* it was exposure to large carnivores on the ground: here there was no peaceable kingdom. As discussed in chapter 3, Africa's array of large carnivores prior to the Ice Age included nearly all of the living species plus a few others now extinct. Today, outside of the rain forest, large herbivores that do not burrow into the ground or climb trees must run just about as fast as the large carnivores that pursue them. This is more than twice as fast as the speed of a modern world-class middle-distance runner. For millions of years, predators and prey of many stripes have been locked in a kind of evolutionary arms race that has eliminated animals on either side that could not keep pace. Large primates are inherently far too slow ever to have entered this competition.

As modern chimpanzees and baboons illustrate, in the presence of large African predators—especially those that hunt in groups under cover of darkness—it is inconceivable that any primate, lacking weapons, fire, and sophisticated social behavior, could survive without climbing trees for everyday protection. In other words, we can see from the shape of its skeleton that *Australopithecus* was a good climber, and we can conclude from its faunal milieu in Africa that it had no alternative but to put its climbing skills to frequent use.

In general, anthropology has devoted much more attention to what our ancestors ate than to what ate *them*. A welcome departure from this tradition has been Charles Brain's analysis of fossils from South African cave deposits. Brain has essentially assigned the australopithecines to the same category of victims as the fossil baboons preserved alongside them. Presumably, the modern human experience has strongly influenced more traditional anthropologists, who, in addressing the ecology of our ancestors, have focused on nutrition rather than the need to avoid fierce predators. We modern humans need sustenance, just as our ancestors did; in fact, many humans still worry about finding their next meal. On the other hand, we are such an imposing species that large carnivores tend to give us a wide berth—taking far fewer human lives than does starvation. Even so, a modern person skeptical about the dangers of carnivores would be in for a fatal surprise if he or she were to join a small group of fellow humans and attempt to survive with them for several nights in the middle of the Serengeti Plain without fire, shelter, or formidable weapons.

One might ask whether the onset of the Ice Age should not have condemned many carnivores to death, thereby reducing the impact of some of the predaceous species on their victims as well. I noted in chapter 3, however, that large African carnivores actually suffered virtually no extinction during the critical interval. We can understand the reason for their good fortune by observing the ecology of the modern species of carnivores, most of which were alive when the Ice Age began. Today, many groups of carnivores occupy woodland areas during the dry season, when large herds of their prey migrate away. These carnivores then shift back to the open savanna during the rainy season, when the prey return. Just as carnivores now migrate seasonally from woodlands to savannas, they presumably undertook a more general long-term shift toward more open terrain when forested areas shrank early in the Ice Age. Unlike antelopes or australopithecines, the carnivores have never depended directly on particular types of vegetation for food or shelter. All they have ever needed is large, meaty prey of one kind or another that they are able to catch.

In chapter 4 I suggested that the inferred habitat change for large carnivores at the onset of the Ice Age altered the ecological dynamics of large African mammals in general. Before the shift populations of herbivores should have been relatively stable. They were spared the heavy mortality that occurs today when the dry season parches so much of Africa. The seasonal suffering of modern herbivores propagates upward through the food web to large carnivores, including lions and hyenas. The carnivores suffer not so much because herbivores die out but because many of them desert the savannas, leaving the carnivores to seek less abundant prey in woodlands that fringe the savannas. Before the Ice Age the forests and woodlands that cloaked broad regions of Africa should have supported relatively stable, permanent faunas of antelopes and other large herbivores. These stable populations should, in turn, have sustained stable populations of large carnivores—creatures that would quickly have driven *Australopithecus* to extinction with year-round predation had it failed to find refuge in trees.

Testing Hypotheses

Science moves forward through the testing of hypotheses. The evidence that *Australopithecus* should have birthed infants mature enough to cling, whereas *Homo rudolfensis* had not only a large brain but also immature infants, amounts to a test which reveals patterns in accord with the terrestrial imperative: *Australopithecus* could have used its evident climbing skills while its infants held onto their mothers. On the other hand, *Homo rudolfensis* has the look of a totally terrestrial creature and could not have climbed habitually anyway, because it had to tote helpless infants.

Fortunately, it has also been possible to conduct a test of the idea of a catastrophic origin for *Homo*, at least indirectly. We can predict that, if climatic changes did indeed trigger the transition to *Homo*, then certain other groups of large African animals should have changed in similar ways. The antelopes are such animals, and chapter 7 explained how we can turn their history into a test.

As it turns out, the fossil fauna of African antelopes, which includes numerous species for the critical interval, displays the same general pattern as that of hominids across most of the continent. Elisabeth Vrba has shown that, close to the time when the Ice Age began, the antelope fauna of Africa changed dramatically. Species adapted to forested habitats died out in large numbers, and new species adapted to open, grassy terrain appeared. Many large four-legged mammals that depended on forests for food or refuge apparently were doomed by the cooling and drying of climates that favored grasslands. Our own family was still restricted to Africa. Why, we can fairly ask, should its members, who depended on forests, not have responded to the widespread environmental changes in much the same way as the kinds of antelopes that also depended on forests?

The Element of Chance

The transition from *Australopithecus* to the human genus was part of a biotic revolution that restructured the African ecosystem in a way that has never been reversed. Revolutions of this type are not scripted in advance. I have emphasized that the appearance of *Homo* on earth was utterly unpredictable. If intelligent extraterrestrial observers, expert in the study of organic evolution, had scrutinized the australopithecines for a million years before *Homo* began to evolve, they would have seen no clues as to what was soon to happen within the human family.

This is not to say that evolution is always totally unpredictable. Once natural selection has begun to move in a particular direction under the obvious influence of some environmental change, then it may be a good bet that the trend will continue. There is no better illustration of this than the celebrated example of so-called industrial melanism in England. Long before modern humans had reached the British Isles, a common species of moth had evolved a speckled black-and-gray pattern that provided camouflage against the bark of the trees on which it habitually rested. The industrial revolution upset this adaptation, darkening the trees with soot and leaving the moths conspicuous to predators. A small percentage of moths had always been black rather than speckled, however, and natural selection suddenly began to favor them. An observor watching this evolution get under way could have predicted that black would become the predominant color in a few years, as indeed it did. Also predictable was the shift back toward the speckled pattern that occurred in many regions when emissions of black smoke declined during the twentieth century. Until an environmental shift that may influence evolution is in progress, however, all predictions are off, and the onset of the Ice Age was unforeseeable during the long period when *Australopithecus* flourished. Suddenly, with the onset of the Ice Age, the game of natural selection changed.

The limitations of evolutionary prediction extend beyond the vagaries of environmental change. Even if imaginary evolutionists from outer space had anticipated the coming Ice Age, they could never have foreseen the evolution of *Homo*. They could not have anticipated the genetic changes that happened to provide natural selection with the raw material needed to grow the large brain of *Homo*. Even retroactively, we cannot assess what the odds would have been, two and a half million years ago, for the chance occurrence of the sorts of mutations that natural selection required to expand and restructure our ancestors' brains before these animals became extinct. (Industrial melanism was much simpler, entailing nothing more than changes in the relative abundance of two color patterns that had been present all along.) For all we know, the evolution of a creature fitting the general description of *Homo* was extremely improbable even after the Ice Age had begun and *Australopithecus* was struggling to survive. Perhaps only a handful of individuals beat long odds in making the momentous transition.

When we seek the ultimate cause for the evolution of *Homo*, we must look beyond events in Africa to the cause of the Ice Age itself. I described in chapter 3 how the Isthmus of Panama rose out of the ocean shortly before the Ice Age began, changing oceanographic conditions in the Atlantic Ocean and virtually shutting down the flow of temperate North Atlantic water into the Arctic Ocean. Once isolated in this way, the polar ocean became the frigid Arctic pond that has existed ever since. This Arctic pond immediately began to refrigerate the surrounding land and ocean, and cold masses of air and ocean water brought cooler temperatures to lower latitudes. Cooler air masses over oceans carried less water to nearby land areas, including central Africa. Thus, a chain of events propagated from the Panamanian region northward through the Arctic region and southward again to the tropics. It is indeed mind-boggling to contemplate the notion that movements deep within the earth, by uplifting a narrow neck of land between two vast oceans, ultimately led to the evolution of the human genus in a land many thousands of miles away.

The Demise of Simple Gradualism

The idea that *Homo* evolved only because a global climatic change disrupted the comfortable life of *Australopithecus* and redirected human evolution conflicts sharply with the earlier belief that for millions of years human evolution moved unrelentingly from an apelike ancestor toward *Homo sapiens*. The prominent role of chance factors in the catastrophic origin, as I have described it, runs counter to this anthropocentric, or human-centered, view.

A few years ago it seemed possible, by a simple connect-the-dots exercise, to incorporate all known fossil remains of the human family, except those of the robust australopithecine *Paranthropus*, into a single, gradually evolving lineage. The resulting graphic, variations of which appeared in numerous books and articles, depicted evolution as molding *Australopithecus* into the modern human species by way of a form of early *Homo* and then of *Homo erectus*. A typical family tree consisted of a single stem, with just one stubby branch jutting off to the side representing the ill-fated *Paranthropus*. Some versions made the Neanderthals a separate branch as well, but usually one that constituted a subspecies of *Homo sapiens* rather than a distinct species. In these highly linear reconstructions the single-species hypothesis often served more as a governing principle than a mere hypothesis. Only *Paranthropus*, it seemed, might have lived alongside our ancestors as an aberrant, giant-jawed creature whose distinctive way of life kept it out of the way of *Australopithecus* and *Homo*. During the past few years new discoveries have revealed the bushier configuration of our actual family tree—overlaps of various species in time—and have pointed to rapid origins for both our genus and our species.

One idea that automatically disintegrated with the bushing out of our family tree was referred to in chapter 1 as the up-from-the-apes view of our origins—the view that over millions of years our ancestors gradually rose from knuckle walking to an upright posture on the ground. In fact, we can now see that no such postural change occurred, either gradually or suddenly. The ancestors of *Australopithecus* were not quadrumanous animals that traversed tree branches on all fours like modern apes and then descended to the ground in this same posture to become knuckle walkers. The age-old assumption that our ancestors elevated themselves from knuckle walking to move about on two legs obviously grew out of the natural tendency of scientists to employ living apes as models of our ancestors. Chimpanzees, gorillas, orangutans, or gibbons are the only species of apes that we can study in the flesh, but we evolved from apes of a different kind.

The early idea that a single path of evolution led inexorably to a modern human pinnacle was perhaps the most dignified picture of our origin that could be salvaged after Darwinism displaced creationism. The mortifying news that our early ancestors were apes continued to unsettle intellectuals long after the famous debate of 1860, in which T. H. Huxley, known as Darwin's "bulldog," excoriated the Anglican bishop Samuel Wilberforce for challenging him as to "whether the ape in question was on your grandfather's or your grandmother's side." A desire to distinguish us from apes seems to have motivated anthropologists of the early twentieth century, such as Arthur Smith Woodward and Arthur Keith, when they insisted, without evidence, that expansion of the brain had begun long ago and led the way in human evolution. On the other side of the Atlantic, Henry Fairfield Osborn, director of New York's American Museum of Natural History asserted, also without evidence, that the human line of evolution had departed from that of apes very early in the Age of Mammals—that we are far removed from such lowly creatures. These anthropocentrists would be doubly insulted by the modern picture, which has the human line of evolution departing from a group of apes quite late in the Age of Mammals and then moving only by fits and starts in the direction of *Homo sapiens*.

Stability and Punctuation

In general a punctuational pattern of evolution entails rapid origins of species followed by long intervals of relatively minor change. In recent years, as new fossil teeth and bones have surfaced, the human family tree has come to look more and more punctuational. Let us review the evidence.

Although not identical, *Australopithecus afarensis* and *Australopithecus africanus* closely resemble one another, especially when compared to the long-legged, big-brained *Homo rudolfensis*, which seems to have evolved abruptly from *Australopithecus africanus* in southern Africa. Taken together, the two species of *Australopithecus* illustrate the survival of a particular form of life for about a million and a half years with little change. Recall, too, that the known remains of *Homo rudolfensis* closely resemble those of *Homo erectus* and that *Homo erectus* itself seems to have undergone relatively little transformation from its time of origin in Africa, about 1.8 million years ago, to the demise of the last of its populations, between 250,000 and 150,000 years ago, in China. At least three or four hundred thousand years before its extinction, *Homo erectus* gave rise to *Homo neanderthalensis* (which some workers call archaic *Homo sapiens* even though it may not be our ancestor); this transition probably took place in Africa, where the Bodo skull of *Homo neanderthalensis* dates back to about 600,000 years. Then, more than half a million years later, Neanderthals overlapped in time with *Homo sapiens* in the Middle East and Europe.

When one examines gradual evolutionary change within any of these lineages, it is easy to lose sight of the forest for the trees, focusing on a bit of change without recognizing that it is just that—a *bit* of change. To assess whether the gradual evolution of an entire species is truly significant, we need a yardstick. For the human family the appropriate yardstick is the distance that any species moved in the direction of a descendant species over a long span of time. In other words, it is not enough simply to note the *presence* of gradual evolution; we also must evaluate the *trajectory* of this evolution. If the evolution failed to move a species very far toward a known descendant species, then we must contemplate a rapid step for the eventual transition—a punctuational origin for the new species. When we move from the species of *Australopithecus* to those of *Homo* in such considerations, the quality of our yardstick improves, for it is when we compare fossils of these species to their close relative *Homo sapiens* that the full range of the variability of modern humans is at our disposal.

Variability is, in fact, a hallmark of our species. Africa alone harbors the Efe, a Pygmy people whose height averages under 140 centimeters (4 1/2 feet), but also the Nilotics, whose males are nearly 40 percent taller than this and have much longer legs for their heights. Other variations extend beyond the obvious physical differences among the major human races. Relatively heavy brow ridges even adorn some modern populations, such as the Aborigines of northern Australia and the Ainu populations of islands that border East Asia. This great diversity serves notice that extinct species of the human genus may have varied substantially in form from population to population. When just a few fossil remains, scattered through time, represent a species, we must avoid leaping to the conclusion that any differences between them necessarily constitute substantial evolution. Much of the observed variation may have existed throughout the species' existence for hundreds of thousands, or even millions, of years, and yet we may have uncovered little of this variation because the fossil record is far from complete. Furthermore, we should expect races with unique features to have cropped up now and again within the lifetime of any species. This is my justification for agreeing with those who accept *Homo neanderthalensis* as a single species, not only the classic European Neanderthals but also such forms as the Atapuerca skulls of Spain, the so-called Rhodesian man and Heidelberg man, and the Solo population of Java. In this scheme the classic Neanderthals were simply a late European race belonging to a wide-ranging species that had a bigger brain than *Homo erectus*.

The hominid fossil record is not complete enough to apply the yardstick for gradual change definitively to every lineage, but there are key segments of our family tree that offer the opportunity. *Homo sapiens* evolved in Africa from either *Homo erectus* or some descendant of it. There has been strong, sometimes acrimonious, debate as to the degree of gradual evolution that *Homo erectus* and *Homo neanderthalensis* experienced. This controversy has obscured the fact that these species shared many features—robust skeletons, for example, and also massive brow ridges, prognathous jaws, and flattened skulls—that disappeared quite rapidly, geo-

logically speaking, in whatever population became *Homo sapiens.*
Neither of these two species of *Homo,* either of which may have
given rise to our species, experienced significant modification of
these features in our direction during a long stay on earth. In addition, there is little evidence of a substantial increase in cranial
capacity within either this species or *Homo erectus* or *Homo neanderthalensis* before the origin of *Homo sapiens.*

Punctuational Evolution of Culture

Support for the punctuational view of human evolution also
comes from another remarkable pattern that, although widely recognized, has been largely ignored by proponents of gradualism.
This is the punctuational evolution of stone tool culture — evolution that generally moved a step behind the evolution of skeletal
features. *Homo rudolfensis,* then *Homo erectus,* and finally *Homo
neanderthalensis* each maintained a distinctive and nearly stagnant
stone tool culture that appeared rather abruptly some time after
the origin of its maker. Furthermore, just as each successive maker
had a larger, and presumably more advanced, brain than its predecessor, each industry was more sophisticated than the one before. Early in its existence, between about 1.8 and 1.4 million years,
Homo erectus stuck with the simple Oldowan stone culture that it
had inherited from its forebear, *Homo rudolfensis.* It then invented
the more advanced Acheulean technology. After the Acheulean
came the Mousterian culture of *Homo neanderthalensis,* which
had spread throughout Africa by about 200,000 years ago but did
not replace the Acheulean instantaneously; occupants of France,
for example, were still crafting Acheulean artifacts as late as 130,000
years ago. We can easily understand the failure of stone cultures
to advance simultaneously in all regions by considering the varied
levels of technology displayed as recently as the nineteenth cen-

tury, when some human cultures remained stagnated at a Stone Age level.

There also is some evidence of early technology transfer. The bones and Upper Paleolithic artifacts of *Homo sapiens* first appear in Europe in strata close to 35,000 years old. It is interesting that from Saint-Césaire, France, there has come a Neanderthal skeleton preserved with a mixture of Mousterian and Upper Paleolithic stone implements. This is a late Neanderthal dated at 36,000 years, plus or minus 3,000. It is highly improbable that Neanderthals, having remained mired in their Mousterian culture for 200,000 years, on their own suddenly learned how to strike well-formed stone blades. In all likelihood they learned some new tricks from *Homo sapiens*, newly arrived from Africa.

In fact, with *Homo sapiens* everything changed. Our species abruptly broke the pattern of cultural stagnation that had previously characterized successive species of *Homo*. It embarked on a course of gradual cultural evolution that accelerated at an exponential pace to put us where we are today. The unique generative powers that had emerged with the evolution of our species' large brain transformed the nature of cultural evolution. No longer did the origin of an altogether new technology require the origin of a new species.

A Modified Single-Species Hypothesis

The single-species hypothesis—the idea that only one member of the human family has walked the earth at one time—is patently false for our ancestors who retained adaptations for climbing: "*Homo*" *habilis* and species of *Australopithecus* and *Paranthropus*. All of these less brainy species, which probably lived semiarboreal lives, variously overlapped for long stretches of time with each other or with *Homo rudolfensis* or *Homo erectus*. But then the australopithecines had a far simpler behavioral repertoire and a much narrower mode of life than *Homo*. It is not difficult to imagine how *Australopithecus* might have lived side by side in Africa with *Paranthropus* by specializing in different foods. The powerful jaws and huge molars of *Paranthropus* probably adapted it to a diet centered on harsher foods, including roots and tubers. Some workers have inferred from various lines of evidence that *Paranthropus* also favored more open terrain than *Australopithecus*. *Paranthropus* obviously avoided conflict with *Homo erectus* to the degree that the two coexisted in Africa for the better part of a million years. Perhaps the two hominids were fully as compatible as are the forest-dwelling Pygmies and the chimpanzees of modern Africa.

Yet I suggest that the pattern of *later* human evolution that is now emerging favors a modified version of the single-species hypothesis. Within the human family this hypothesis may hold only for *Homo*—and for this genus only in the sense that its species seem to have been able to overlap with one another for but modest intervals of geological time.

It appears that the broad ecological niche that has characterized every species of *Homo* has made it difficult for any two of the species to live together. Their unprecedented intelligence generated ecological versatility. They ate everything from insects to horses and from grains to coconuts. Their ability to plan for the future, pass wisdom from generation to generation, and surmount environmental barriers expanded their ecological role and gave them a towering presence in the environment. These traits, and perhaps a new kind of curiosity, led *Homo erectus* and younger species to invade new continents, so that there was little opportunity for any two contemporary species to remain geographically separated for long.

It was probably very early in its history that *Homo* acquired an ability that has since burgeoned into a scourge that may destroy it. This is the ability to launch physical attacks with manufactured weapons, rather than simply with hands and teeth in the primitive manner of chimpanzees and, presumably, australopithecines. When it is directed at lower animals, we call the more advanced practice "hunting." When it is directed at other members of our species, and perhaps other species of our genus, we refer to it as "murder" or, when politically motivated, "warfare."

Our species likely played a major role in the disappearance of the classic Neanderthals from Europe through some combination of physical attacks and competition for food or shelter, even though the two species had existed for tens of thousands of years in the region of Israel and perhaps for even longer in Africa. Other intervals of overlap between species of *Homo* are even less well defined, but it appears that they may seldom have exceeded one or two hundred thousand years.

Homo *as a New Adaptive Complex*

I believe that a profound, natural demarcation separates what we might call the adaptive complex of true *Homo* (big-brained forms, starting with *Homo rudolfensis*) from the australopithecine adaptive complex, including not only *Australopithecus* and *Paranthropus* but also *"Homo" habilis*, the species that may warrant placement in a new, as yet unnamed, genus of australopithecines. In other words, the advent of true *Homo* ushered in a new phase in human evolution. The salient traits of the australopithecine adaptive complex were these:

- a relatively small brain
- an apelike schedule of postnatal maturation
- semiarboreal adaptations, with limited ability to run on the ground
- a large disparity in size between males and females
- promiscuous sexual behavior rather than pair-bonding

- the ability of species to coexist with other hominid species (violation of the single-species hypothesis)
- the failure to manufacture stone tools

The contrasting adaptive complex of *Homo* includes these traits:

- a very large brain with an expanded prefrontal cortex
- delayed postnatal development compared to that of an ape
- anatomical adaptation to an essentially terrestrial life (some humans still occasionally climb trees)
- a modest difference in size between the two genders (at least in advanced species)
- Pair-bonding between the sexes (at least in advanced species)
- the inability of the two species to coexist for very long (adherence to the single-species hypothesis)
- the fabrication of stone tools or more advanced artifacts

The notion that the human family falls naturally into two quite different adaptive complexes runs counter to the views of those anthropologists who continue to insist that *"Homo" habilis* was a valid member of our genus that included both the small-brained fossil forms generally agreed to represent this species and the big-brained individuals now recognized as *Homo rudolfensis*. This lumping of quite different crania obviously blurs the distinction between the two adaptive complexes that I have delineated. The workers who still unite the quite different crania focus heavily on the teeth of *"Homo" habilis*, which to a degree resemble those of unquestioned *Homo*; these workers, in my view, pay too little attention to the small brain of *"Homo" habilis*, the apelike proportions of its limbs, the absence of advanced balancing structures in its ears, and the tiny size of its females. All of these features have profound implications.

Clearly, human evolution crossed a major threshold when *Homo* emerged about two and a half million years ago, even though "*Homo*" *habilis* and *Paranthropus* lived alongside the new genus for some time thereafter. The new adaptive complex was all of a piece. The delayed development and large brain were, of course, interrelated, and they were incompatible with everyday tree climbing. The advanced brain led not only to tool manufacture but also indirectly to pair-bonding and, hence, to a reduced difference in the average body size between males and females. *Homo* soon put its brain to good use, in part by constructing advanced social structures that could not function if males frequently contested with one another for females. Thus, sexual selection that had favored especially large males came to an end.

Evolution as a Compromise

While developing the theme that human evolution has not followed a simple, linear path from the earliest australopithecines to *Homo sapiens*, I have emphasized the compromises that evolution has negotiated along the way. Many of these are still with us, leaving us far less perfect animals than we might wish to be.

The adaptive compromises embodied in the skeleton of *Australopithecus* are the evolutionary results of living a double life. *Australopithecus* was both a reasonably good climber, so that it could have made good use of copses of trees as protective home bases, and also a reasonably good walker, so that it could have foraged effectively for food on the ground and migrated between home bases efficiently. But, as the product of evolutionary compromise, it could neither match the acrobatics of an ape in a tree nor run as we can on the ground.

For *Homo* I have discussed adaptive compromises of a different kind. They represented the inability of the evolutionary process to perfect even a creature that has committed itself totally to life on the ground. One of these compromises was the evolutionary descent of the voice box. While this change facilitated speech by providing space for movements of the tongue, it prevents us from breathing while we eat and allows food occasionally to block our windpipe. Another compromise was the expansion of the pelvis that permitted both *Homo neanderthalensis* and *Homo sapiens* to give birth to large babies—but only at some loss of mechanical advantage for hip muscles during two-legged locomotion.

Retardation of our development after birth was the most profound trade-off of all, however. In the magnitude of both its positive and negative effects, it exceeded all other compromises in human evolution. By extending the high rate of brain growth that all primates experience before birth, it endowed infants of early *Homo* with a huge brain not long after they were born. (Recall that my calculations show even *Homo rudolfensis* experiencing delayed development much like ours.) One negative result of this giant positive step was energetic: the larger brain required many calories to sustain its activity—literally, food for thought. Much more problematical, however, was the period of time that *Homo* ended up spending in a state of infantile helplessness. (Recall that here, as in our degree of brain expansion immediately after birth, we modern humans rank number one among all living species of mammals.) Since its inception *Homo* has been forced to tend and transport and protect embryolike infants.

In descendants of *Homo erectus*, natural selection extended the interval of rapid brain growth further by the widening of the pelvis, so that the high rate of early brain growth could continue longer inside the womb. One way of viewing the overall extension of the fetal growth pattern is to think of modern humans as having a gestation period of twenty-one months, nine inside the womb plus twelve in the outside world during which the brain continues to grow at the high fetal rate in babies that remain highly immature.

My measurements suggest that *Homo erectus* expanded its brain at the rapid fetal rate by about the same percentage as we modern humans do after birth. Having entered the world with a smaller body and brain, however, it ended up with a smaller body and brain at the stage when the rapid rate of brain growth ceased—a stage that modern human infants reach at an age of about one year. *Homo erectus* was probably pushing juvenilization to the limit, just as our species was likely doing before the advent of medical facilities that nurture premature babies. In fact, the birthing of immature babies must have created major ecological problems for early members of the human genus. Babies substantially more embryonic than ours at birth would likely be intolerable in nature because they would suffer very high rates of infant mortality and would impose an impossible burden of dependency on parents.

The origin of our large brain exemplifies evolutionary opportunism at its best. In effect, natural selection took advantage of a preexisting aspect of development—a fetal pattern of growth—and extended it into the postnatal interval. Presumably all that was required for evolution to start down the path it actually took—the path of delayed development—was for one key mutation—or, at most, a very few—to arise within a population of *Australopithecus* and then spread throughout it. The odds probably had been enormous against natural selection's enlarging the australopithecine brain dramatically by moving along any other path than this one. The path to the big brain that evolution followed was worth the ecological and energetic sacrifices it entailed.

Qualities of the New Brain

We must not overlook the fact that, when evolution created *Homo*, it did not simply enlarge the brain but also changed its structure. The brain grew much more than was required for managing the muscles of a slightly larger body, so that the excess brain tissue was available for expanded cognition. Sectors of the brain devoted to particular thought processes vied for unclaimed neurons by a remarkable sorting process. Early in life, neurons and connecting fibers proliferated rampantly throughout the brain but then withered away if they proved not to be well connected. Only those that were put to good use received sustenance. By this process the neural networks of *Homo* took form. The brain of early *Homo* undoubtedly differed in many ways from that of *Australopithecus*. Unfortunately, the interior surface of a skull records so little of the convoluted topography of the brain that fossils frustrate us with meager information about the details of the reorganization. What we can easily see is that in the evolution of *Homo* the prefrontal cortex of the brain expanded as the forehead became elevated to become a tall, nearly vertical facade for the front of the brain — and the prefrontal cortex is where we think.

Lateralization — the installation of new functions on only the left or right side of the brain — also came into play. When a new function arose on only one side of the brain, then the function that had previously been positioned there survived in the corresponding region on the other side. Expansion of the entire brain during this displacement process then provided more space for the original function on the side where it remained. The evolution of this formidable asymmetrical brain with its massive prefrontal cortex endowed *Homo* with a level of cunning, creativity, and socialization that the world had never seen before.

Would that we could visit with earliest *Homo* to get inside its head. Sadly, we will never even know the extent to which it could communicate with its comrades by sign language or verbal utterances. Even for the more advanced Neanderthals, language is at issue. Because the simple Neanderthal culture may not have required sophisticated communication, Christopher Stringer and Clive Gamble, experts on these animals, have doubted whether they employed advanced syntax. Perhaps language, as we modern humans know it, is our unique preserve. Even so, early *Homo* must have engaged its large brain for advanced communications and social interactions far beyond anything that *Australopithecus* ever achieved.

Aggression and Love

Some portion of *Australopithecus* survives within us—in our genes, in our bodies, and in our behavior. In the 1960s the playwright Robert Ardrey provoked the general public with popular writings that pinned a new kind of original sin on humanity. We are innately vicious, he claimed, the genetic heirs of killer apes. Certainly, we harbor aggressive attitudes and occasionally act on them, but their origins are difficult to discover.

It is true that chimpanzees hunt small game, but this forms only a small percentage of their caloric intake. Male chimpanzees do most of the hunting, perhaps because they are much larger than females. Other species of modern apes appear to consume little or no meat, and there is no reason to believe that an instinct for hunting was ingrained in the particular apes who were our ancestors. For all we know, these animals were strict vegetarians. As for *Australopithecus*, we can only conjecture that it may have supplemented its largely herbivorous diet with some small game.

We have only a sketchy picture of the feeding habits of *Homo* prior to thirty thousand years ago. Because members of the human family must first have begun to hunt by taking small animals without weapons, we will probably never know when this activity began. Earliest *Homo* may have assaulted larger prey with hand-hurled stones, handheld bashers, and wooden lances, but it seems not to have fixed stone points to spears or arrows and may have garnered most of its meat by scavenging. Quite telling is the absence of spear and arrow points in the time of *Homo erectus* and the presence of very few points hafted for attachment to shafts even among the Neanderthals' leavings. A lone wooden spear, its cellulose body preserved by chance, reflects the general failure of Nean-derthals to move into a new era by fashioning stone- or bone-tipped projectiles. Bones of mammals associated with Neanderthal remains suggest that these large-brained hominids took mainly small and medium-sized game even after fifty thousand years ago. They apparently lacked the weaponry to concentrate on elephant, woolly rhinoceros, or bison.

From the time of *Homo rudolfensis* until the emergence of *Homo sapiens*, it would seem that members of the human family relied heavily on intelligence and cooperation to capture prey. They presumably stalked animals, ambushed them, and chased them into bogs or over cliffs. For more than two million years our ancestors designed stone tools primarily for skinning and butcher-ing. They did not generally fashion points well for killing at a dis-tance—for attachment to the tips of projectiles—but only for bashing.

A special interest in hunting, stimulated by the analysis of stone tools, may have caused us to exaggerate the percentage of meat in the diets of the hominids who preceded our species. De-bates about the relative importance of hunting and scavenging have obscured the issue of whether plant foods formed a vastly larger portion of total caloric intake than meat. There can be lit-tle doubt that meat was a staple for the Neanderthals, because their bones so frequently occur with those of the medium-sized animals they obviously butchered. For all we know, however, their imme-diate ancestor, *Homo erectus*, relied much more heavily on plant foods.

If we wish to scrutinize our ancestry for the source of human aggression, we can also look to the kind of territorial behavior that has troops of chimpanzees exterminating each other in a pre-meditated manner unknown among lower animals. But here, too, we can only reason from analogy. Perhaps australopithecines made war in the way that chimps do, but perhaps they did not; chimps were not their ancestors.

We also must bear in mind that, ironically, the expansion of hunting by the human genus, while directing aggression at other species, actually reduced aggression between members of hunting parties. The hunters needed to cooperate in order to succeed, as apparently manifested by the substitution of pair-bonding for competitive polygamy.

The evolution of *Homo* would seem also to have favored the emotion of love within a species. Pair-bonding is at the root of love between the sexes in modern humans. Certainly, natural selection also favored the love of offspring because, for adults of both genders, this form of affection tended to ensure that their progeny were well trained and cared for — that they would be more likely to survive and pass the parents' genes on to later generations.

Our Evolutionary Future

This book has concerned our past. Our modern biological condition and culture are the products of this history. By some standards we are far richer than our recent ancestors, although we have no proof of being any happier. Even if we adopt a yardstick for progress that puts us far ahead of *Homo erectus* or *Homo rudolfensis*, we must acknowledge our humble origins.

From an evolutionary standpoint we are nouveau riche. We might wonder whether biological evolution during the coming millennia may make us even richer — or at least different. Knowledge of our evolution tells us how we have come into being, but does it also disclose where we are going?

I am frequently asked what our evolutionary future may hold. My initial response is to point out that *Homo sapiens* have evolved only in trivial ways for thousands of years; our biological traits, like those of most other well-established species, have remained relatively stable. This is now entirely irrelevant, however. The answer as to where we are going is quite simple: our future is largely in our own hands. The advent of modern medicine disrupted natural selection, introducing *un*natural selection. Today we heal or reconstruct many individuals who would never have survived in the wild. Believing in the sanctity of human life and seeing nearly all people as having something to offer, our culture sustains individuals who have severe physical or mental defects and permits many of them to reproduce.

In 1981 I discussed our shift to unnatural selection in a book called *The New Evolutionary Timetable*, and I also noted that we have limited potential to give rise, unintentionally, to any new species. In fact, even before advanced medicine and other aspects of modern culture interfered with natural selection, there was little chance that our species, once it had spread around the world, would have given rise to any additional species outside of the laboratory. Very few local human populations have remained so completely isolated on our planet as to have had the opportunity to diverge dramatically in a new evolutionary direction. If a new species is to arise from ours without intentional genetic tinkering on our part, the most likely venue is outer space. If we eventually scatter small colonies of humans outside our solar system, one of these might conceivably reach a state of reproductive isolation by evolving endemic genetic changes in the long absence of contact with earthlings.

Here on earth, even now, there is little doubt that we could create a new species from a small population of *Homo sapiens* if we wished. It would presumably be impossible for us to transform our entire species into something quite different, but, through artificial breeding or artificial insemination of a captive population, we could create a divergent class of beings able to breed successfully with each other but not with us. We could do this in much the same way that Native Americans long ago created Chihuahuas from wolves, and genetic engineering endows us with powers much greater than those of traditional animal breeders.

Such fantasies raise the ethical issues associated with eugenics, a field that traces back to the nineteenth century, before the birth of modern genetics. Francis Galton, the biologist who coined the term "eugenics," believed it to be the duty of science to better the lot of humans not simply by ameliorating the environment but also by improving the species itself. Eugenics retained widespread academic and political respectability well into the twentieth century. Early on, however, it came to smack of racism. For example, the restrictive Immigration Act of 1924 swept through both houses of the United States Congress, and President Calvin Coolidge signed it with enthusiasm. The bill restricted the flow of immigrants to small fractions of the particular numbers of individuals who had come from various foreign countries in 1890, a time when few people had arrived from southern and eastern Europe.

The emphasis in eugenics later shifted from favoritism for certain races to artificial selection of certain individuals—individuals deemed superior to others. Eventually, beginning in the 1930s, Nazi Germany gave eugenics its enduring bad name. Adolf Hitler's program began with forced sterilization and, of course, descended to genocide.

Some of the ways in which we can alter our biology are encumbered by few ethical issues. Gene therapy, for example, resembles other medical procedures. Techniques that will allow us to design our own offspring are another matter. In a few years the Human Genome Project, which engages thousands of researchers, will produce a complete road map of our chromosomes, marking the sites of genes that perform various functions.

Putting the map to use will not be quite as easy as it might appear. Commonly, many genes affect a particular feature of an animal. On the other side of the coin, a particular gene often affects several features. In addition, genes interact with one another, and some govern the operation of others, turning them on or off. Although we will never find it easy to restructure ourselves along the lines of some grand design, the possibility is now before us.

Aging is one aspect of our biology that we may legitimately

choose to attack. It is tantalizing that some animals age almost imperceptibly. Perhaps genetic engineering can substantially extend our biblical allotment of three score and ten years, which is already about a decade longer than the natural longevity of a chimpanzee. How will we treat such power?

Far more perplexing will be our eventual ability to expand human intelligence. Even today, we could increase average intelligence by selective breeding. Humans exhibit a wide range of innate mental abilities. At age fourteen, for example, Mozart listened to the *Miserere* of Gregorio Allegri in the Sistine Chapel and, afterward, set the entire body of music down on paper from memory. John Stuart Mill began learning Greek at the age of three and soon began studying the classics. To amuse themselves during a train ride, the nuclear physicists Edward Teller and Otto Frisch once played chess without a board, using only mental images. As the movie *Rain Man* depicted, there are people in the world who, although severely retarded in many other ways, can tell you at a moment's notice the day of the week on which any date in the next century will fall. The particulars of IQ tests show that many people are brilliant in some components of intelligence but average or even subnormal in others. No one would suggest that such differences are entirely the result of environmental factors. Such discrepancies within individual people, along with the left-brain–right-brain dichotomy and other evidence of localized brain functions, indicate that there are multiple components of intelligence in humans and that to some degree these are under separate genetic controls.

There is a sperm bank in the United States that contains only deposits from purported geniuses. No one has embarked on a program to put such a reservoir to use over a period of several generations, but the bank would not exist in the absence of the intent to increase intelligence in some population of humans through selective breeding.

Genetic engineering will add a dimension to our ability to impose artificial selection on basic intelligence. By introducing new genetic components, it will open the way for us to elevate our maximum brainpower, if we so choose. It will someday give humans the potential to create "designer children" with high levels of intelligence or other traits that seem desirable. *Homo sapiens* will be able to endow some of its members with remarkable abilities never seen before. If we ever embark on such an audacious and potentially perilous course, we might accelerate our own evolution to a rate faster than any punctuational transition in the history of the human family.

As we acquire greater and greater power to manage our evolution, what legal and ethical limits will we impose upon ourselves? So dramatically is genetic engineering now promising to expand our power to dictate our biological future that we seem unable to acknowledge the ethical quagmire that it places in our path. How often do we hear the thorny issues of self-directed evolution discussed in detail?

Unnatural Selection Is Risky

Ethics are not our only concern. What we know of the human evolution that has already occurred points to problems of a more practical nature in any attempt to alter our species drastically. In general these problems are related to the difficulty of modifying a marvelously intricate self-regulating system: a living species.

Before a climatic change allowed *Homo* to evolve, natural selection failed to expand the brain of *Australopithecus* appreciably despite having worked in what amounted to a vast natural laboratory for more that a million and a half years. Human research is confined to much smaller laboratories and much shorter spans of time. While we are nonetheless developing powerful abilities to manipulate our genetic architecture, any effort to reinvent ourselves will confront obstacles.

In particular, the compromises made by natural selection in the past serve a warning. A genetic change that is beneficial in one way often is deleterious in another. Most notably, of course, the change or changes that expanded the brain of *Australopithecus* into the brain of *Homo* saddled members of our genus with infantile helplessness. Natural selection cannot tinker with species easily. By their very existence species reveal themselves to be biological entities that have functioned well. The transformation of such a living being into something quite different that also functions well is a wrenching event. Piecemeal alteration of a species by genetic engineering is a risky proposition because many genes affect several biological features of a species in ways that may be unrecognized at the outset. Artificial genetic alteration that solves one problem may create others that are even more severe. Natural selection does not behave in this way, at least if we assess the net value of the changes that it brings about in terms of longevity and fecundity. Whatever compromises it makes, it tends to favor individuals that leave behind relatively many fertile offspring. Such individuals tend to be healthy and long-lived. On the other hand, genetic engineering on a large scale is bound to yield unpredicted side effects that are cruelly inimical to its live products.

Turning Against Our Natural Habitat

The problems that we humans now confront are infinitely more complex, if not more terrifying, than those the immediate ancestors of our genus faced when a changing climate flushed them out of the trees. These transitional australopithecines were the passive victims of an environmental revolution. We modern humans may take heart that we are less at the mercy of our habitat. On the other hand, our brain, half again as large as that of early *Homo*, threatens us with runaway cultural evolution; we face perplexing environmental problems that issue from our own manipulation of nature. While we have begun to influence our own evolutionary fate by interfering with natural selection, we have also been reshaping the environment that once guided our evolution. Unfortunately, many of our actions degrade our habitat because we undertake them in order to reach goals whose allure blinds us to myriad dire consequences.

The impact of human civilization on the earth's climate is profoundly ironic in light of what this book has shown about our origins. The Ice Age birth of our genus must convince us that not all global climatic change is inimical, but we may soon find ourselves reacting emotionally to changes of our own making in much the way that *Australopithecus* must have felt when its protective forests shrank away. It turns out that in order to fuel our complex civilization we are lacing our planet's atmosphere with carbon dioxide, a greenhouse gas that, if it has not already begun doing so, will soon warm the Ice Age climate to which we owe our very existence.

Thresholds

Our evolutionary origin was a fortuitous result of one small geological event, and it carried no guarantee of long-term success. Paradoxically, while chance events played a major role in the successful emergence of *Homo*, now that we can influence both our environment and our evolution and thus minimize the impact of fortuitous external events, our future is very much in question. We manipulate our habitat with no consistent rationale, even as we confront vast new powers to mold the bodies and brains of those to whom we will bequeath the world of our making.

It all comes down to thresholds. When environmental change proceeds unchecked, at some threshold a population — or a species or a whole community of species — crashes. If we ignore the intricacy of the global ecosystem and continue to degrade and dismantle various portions of it, then we and our habitat will descend by abrupt steps over thresholds that lead to successively lower levels of environmental quality.

The human genus parallels the whole of nature, possessing its own special intricacy—and unique fragility—that natural selection built into our ancestors as it impelled them across a problematical environmental threshold. If we ignore the evolutionary origin of this particular complexity and allow our brain to turn against the process of natural selection that produced it, undoing some of its masterful work, we may inflict damage on future members of our fragile species quite abruptly and in unforeseen ways.

Our civilization now confronts a related threshold that is cultural in nature—a matter of conscious decision. This is actually a double threshold, entailing our behavior in the domains of both human and environmental change. We have thrown it up without any overarching design by simply moving enthusiastically ahead with ever accelerating science and technology. Standing directly before us, this imposing cultural threshold demands that we choose among a multitude of future biologies and future habitats. In this confrontation even a creative scientist can justly turn reactionary. Let us hope that our great brain can engage in ample self-regulation, reining in some of its potential to transform itself and its habitat when wisdom dictates. Otherwise, the human genus may orchestrate its own downfall two and a half million years after blind chance allowed it to emerge in a perilous, thinly wooded African Eden.

NOTES

5 "Light will be thrown" Darwin, 1859, p. 488.
Darwin, 1871.
For the original formulation of the single-species hypothesis, see Mayr, 1950.

7 "more and more erect" Darwin, 1871, p. 143.
"gradually converted" Ibid., p. 142.
"The evidence now available" Dobzhansky, 1962, p. 220.
"is one of the distinguishing marks" Ibid., p. 221.
For early ideas on the punctuational model, see Eldredge, 1971; Eldredge and Gould, 1972; Stanley, 1975, 1979.

9 "book that favored the punctuational model" Stanley, 1979.
For the evolutionary history of chimpanzees, see Morin et al., 1994.

10 For the first description of the gorilla, see Savage and Wyman, 1847.

12 For an account of the discovery and early interpretation of Lucy, see Johanson and Edey, 1981.

14 For Ice Age mammals, see Kurtén, 1968; Kurtén and Anderson, 1980.

16 "*Paleobiology* published my article" Stanley, 1992.

17 Howells, 1993.

19 For a review that illustrates the traditional lack of emphasis on predation upon human ancestors, see Foley, 1987; discussion of predation is confined to p. 183.

22 For the evolution of mussels, see Stanley, 1972.

23 For the Ice Age extinction of mollusks, see Stanley, 1986.

24 For the transformation of the African antelope fauna, see Vrba, 1974, 1975.

26 For an account of the discovery of "Java man," see Theunissen, 1989. Keith, 1925.

28 Woodward, 1925; Spencer, 1990.

32 For recent information on the ages of important hominid-bearing strata in the Turkana basin, Olduvai Gorge, and the Hadar region of Ethiopia, see Feibel et al., 1989; Walter et al., 1991; Walter, 1994.

For the geology of Olduvai Gorge, see Hay, 1976.

34 For the age of the Taung skull, see McKee, 1993.

35 White et al., 1994.

Leakey et al., 1995.

36 For the sources of the data on hominid brain size, see Stanley, 1992.

For a useful summary of the form and occurrence of the two species of *Australopithecus*, see Klein, 1989.

38 For weights of males and females, see McHenry, 1991b.

For early ideas on sexual selection, see Darwin, 1859, pp. 87–90; 1871.

40 For the early description of the pelvis of *Australopithecus*, see Dart, 1949.

41 For an evaluation of the australopithecine footprints and their geological and paleontological context, see Leakey and Harris, 1987.

As an example of the view that *Australopithecus* was a totally terrestrial biped, see Lovejoy, 1981.

42 Stern and Susman, 1983; Susman, Stern, and Jungers, 1984.

44 Clarke and Tobias, 1995.

45 For an analysis of the mechanics of vertical climbing in primates, see Cartmill, 1974.

46 For the climbing abilities of humans, see Skeat and Blagden, 1906 (Sakei and Semang tribes); Wood-Jones, 1964 (seagoing Malays).

49 For an illustration of the toe-off problem, see Stanley, 1992.

Berge, 1994.

50 For an analysis of the inner ear and locomotory balance of *Australopithecus*, see Spoor, Wood, and Zonneveld, 1994.

51 A group of articles portrays *Paranthropus* in Grine, 1988.

53 Romer, 1959; Rodman and McHenry, 1980.

Lyell, 1833.

54 It was previously believed that the Mediterranean dried up altogether; see Hsü, 1972.

55 For the behavioral ecology of chimpanzees, see Ghiglieri, 1984; Goodall, 1986; McGrew, 1992.

57 For the locomotory efficiency of chimpanzees on the ground, see Rodman and McHenry, 1980.

For the limitations of gripping in *Australopithecus*, see Marzke, 1983; McHenry, 1983.

60 "By 1929 . . . I was claiming" Dart, 1959, p. 104; this book reviews Dart's earlier publications.

61 For the discovery of the pelvis, see Dart, 1949.

"was the first step" Dart, 1959, p. 195.

64 Vrba, 1980.

Berger and Clark, 1995.

66 For a summary of occurrences of hyenas and other mammals in Britain during the Ice Age, see Sutcliff, 1985.

67 For articles on the ecology of the Serengeti, see Sinclair and Norton-Griffiths, 1979.

70 Facts about African carnivores in chapters 3 and 4 are largely from Kruuk, 1972; Schaller, 1972; Bertram, 1979.

76 For carnivores that coexisted with early hominids, see Brain, 1981; Turner, 1990.

For the killing habits of saber-toothed cats, see Ackersten, 1985; Marean, 1989; Van Valkenburgh, Teaford, and Walker, 1990.

78 For the running speeds of large African mammals, see Bertram, 1979; Shipman and Walker, 1989.

80 For the behavioral ecology of baboons, see Devore and Washburn, 1963; Altman and Altman, 1970.

83 Stern and Susman, 1983; Susman, Stern, and Jungers, 1984.

84 For gorilla nesting habits, see Schaller, 1963.

86 For reviews of past ice ages, see John, 1979; Frakes, Francis, and Syktus, 1992.

For a review of extinction during ice ages, see Stanley, 1987.

88 For details of the extinctions at the end of the Eocene epoch, see Prothero, 1994.

90 For a general review of terrestrial biotic changes during the latter half of the Age of Mammals, see Webb and Opdyke, 1995.

91 For a review of climatic and biotic changes during the Pliocene, see Stanley and Ruddiman, 1995.

94 For the history of ideas about the earth's rotation, oxygen isotopes, and glacial maxima and minima, see Imbrie and Imbrie, 1979.

97 For accounts of early debates over the concept of continental glaciation, see Imbrie and Imbrie, 1979; Nilsson, 1983.

101 For modern summaries of conditions throughout the world during glacial maxima, see Nilsson, 1983; Dawson, 1992.

102 For the geology of the Key Largo Limestone, see Stanley, 1966.

103 Imbrie and Imbrie, 1979, summarizes the results of CLIMAP; for technical details, see Cline and Hays, 1976.

104 Distributions of forest and savanna in tropical Africa are compared for the present day and glacial maxima in Hopkins, 1974; Hamilton, 1982; Whitmore, 1990.

107 For a review of the spread of active dunes in Africa during glacial maxima, see Sarnthein, 1978.

Movements of glaciers on Mount Kilimanjaro are reviewed by Ormaston, 1989.

109 Hamilton, 1982, reviews biotic evidence for Ice Age fragmentation of African rain forests.

111 Increased delivery of sand to the Atlantic seafloor during glacial expansion is discussed by de Menocal, Ruddiman, and Pokras, 1993.

For fossil pollen and East African climates, see Bonnefille, 1976, 1983.

112 The transitional woodlands of Africa are described by Hopkins, 1974.

113 For carbon isotopes and ancient floras in Africa, see Cerling, 1992.

117 For carnivores that coexisted with early hominids, see Brain, 1981; Turner, 1990.

118 For chimpanzee reproduction, see Goodall, 1986.

122 For the biology of *Paranthropus*, see Grine, 1988.

125 The intricate structure of Aristotle's lantern is described by Hyman, 1955.

For accessible introductions to the structure and function of the human brain, see Ackerman, 1992; Posner and Raichle, 1994.

126 For the arithmetic of neuronal connections, see Deacon, 1990.

129 For chimpanzee sign language, see Savage-Rumbaugh, 1986; Gardner and Gardner, 1969.

130 For human generativity, see Corballis, 1991.

For the mimetic skills of humans, see Donald, 1991.

134 For neuronal maps, see Edelman, 1992.

135 For environmental stimulation and neuronal networks, see Diamond, 1988.

139 For revelations from PET scans about the involvement of several regions of the brain in the comprehension and production of speech, see Posner and Raichle, 1994, pp. 112–25.

140 Of the many books written on brain asymmetry, two are especially readable: Springer and Deutsch, 1993, and Bradshaw and Rogers, 1993.

Corballis, 1991, pp. 182–85.

142 MacNeilage et al., 1987.

143 The idea that the positioning of a function on both sides of the brain wastes brain tissue has been expressed by Eccles, 1989, and reiterated by Ackerman, 1992, p. 28.

144 For recent results of brain scanning, see Posner and Raichle, 1994.

147 For a pessimistic view of sulcus recognition on the Taung filling, see Tobias, 1987.

For the brain structure of *Australopithecus*, see Holloway, 1983.

148 For the principle of proper mass, see Jerison, 1973.

150 For a brain model based on the principle of proper mass, see Deacon, 1990.

154 For data on the 10 percent rule, see Holt et al., 1983.

For a comparison of Phase I and Phase II growth in humans and other mammals, see Count, 1947.

156 For a review of delayed development and early brain expansion in humans, see Gould, 1977.

158 Calculations and sources of data for patterns of development in *Australopithecus* and species of *Homo* can be found in Stanley, 1992.

164 For the estimation of stature from the diameter of the hip socket, see McHenry, 1991a.

166 For the ages of fossils from the Lake Turkana region, see Feibel et al., 1989.

For a detailed analysis of the Oldowan stone culture, see Potts, 1988.

167 For a description of the jaw of *Homo* that is about 2.4 million years old, see Schrenk, 1993.

For the idea that a behavioral shift often precedes a major change in body form, see Mayr, 1963.

169 Sileshi Semaw, a doctoral student at Rutgers University, announced the discovery of the oldest stone tools at the 1995 meeting of the American Association of Physical Anthropology; see *New York Times*, April 25, 1995, p. C1.

170 For changes in the rodent faunas of Africa with the onset of the Ice Age, see Wesselman, 1984.

171 Her hypotheses are summarized in Vrba, Denton, and Prentice, 1989.

173 For hypotheses as to how delayed development might somehow benefit humans, see Montagu, 1989.

174 For brain metabolism, see Martin, 1989.

176 For his views on sexual selection, see Darwin, 1859, pp. 87–90; 1871.

For the relationship between relative male and female body sizes and pair-bonding, see Clutton-Brock, 1985.

178 The controversy over hunting versus scavenging by the Oldowan toolmakers is illustrated by the contrasting views of Shipman, 1986, and Bunn and Knoll, 1986.

The size of thighbone of *Homo rudolfensis* is from Kennedy, 1983; the data for modern thighbones are from Tague, 1989.

179 For a review of the possible causes of the glacial expansion, see Stanley and Ruddiman, 1955.

For previous ideas on the role of the Isthmus of Panama and other factors in the origin of glaciations, see Frakes, Francis, and Syktus, 1970.

180 For the migration of mollusks across the Arctic, see Durham and McNeil, 1967.

181 For more details of the new explanation for the modern Ice Age, see Stanley, 1995.

182 For the operation of the transoceanic conveyor belt, see Broecker, 1987. There, however, the northward flow of the conveyor belt was positioned in the eastern Atlantic.

184 For the age of the isthmus, see Coates et al., 1992.

188 For a detailed description of the Turkana boy, see Walker and Leakey, 1993.

192 For human body proportions and climate, see Ruff, 1991.

193 For the cranial capacity of Homo erectus, see Holloway, 1983.

195 For the first description of the remains of Homo habilis, see Leakey, Tobias, and Napier, 1964.

196 Exemplifying the view that the small-brained forms were distinct from what is now called Homo rudolfensis were Walker and Leakey, 1978. The view that the large and small-brained forms were, respectively, males and females of Homo habilis was expressed by Tobias, 1987.

For the initial description and discussion of the postcranial bones of Homo habilis, see Johanson, 1987; Johanson and Shreeve, 1989.

198 For a description of the postcranial bones of an apparent male Homo habilis, see Leakey et al., 1989.

For early dates for Homo erectus in the Caucasus and Indonesia, see Gabunia and Vekua, 1995; Swisher et al., 1994.

199 For the history of the discovery of Homo erectus in China and Java, see Lanpo and Weiwen, 1990; Theunissen, 1989.

200 For a review of the fossil record and the general evolutionary stagnation of Homo erectus, see Rightmire, 1990.

201 For the brain size of OH 9, see Walker and Leakey, 1978, figure 14.28.

For the fossils from the Caucasus, see Gabunia and Vekua, 1995.

202 For a review of Acheulean tools, see Klein, 1989, pp. 206–16.

For the early use of fire, see Gowlett et al., 1981.

203 For a discussion of alternative classifications that favor the assignment of so-called archaic Homo sapiens to Homo neanderthalensis, see Stringer, 1993.

For the age of the Bodo skull, see Clark et al., 1994.

For Spanish fossils 780,200 years old, see Carbonell et al., 1995.

For early Neanderthals in Spain, see Arsuaga et al., 1993.

204 For details and references on the classic Neanderthals, see Stringer and Gamble, 1993; Trinkaus and Shipman, 1993.

208 The early African artifacts of Homo sapiens have been described by Yellen et al., 1995.

208 For a review of the history of the controversy over Eve, see Lewin, 1993, chapter 4.

For a summary of the genetic evidence favoring an African origin for *Homo sapiens*, see Cavalli-Sforza, 1991.

209 Occurrences of early modern humans and Neanderthals in the Middle East are reviewed by Stringer and Gamble, 1993, chapter 5.

211 For a review of Cro-Magnon artifacts, see Lewin, 1993, chapters 5–6.

219 Stern and Susman, 1983; Susman, Stern, and Jungers, 1984.

220 Brain, 1981.

223 Vrba, 1974, 1975.

For industrial melanism in moths, see Bishop and Cook, 1980.

227 For accounts of the debate with Wilberforce, see Huxley, 1900. Osborn, 1910.

228 For a review of these various groups of modern humans, see Kingdon, 1993.

229 For a broad definition of *Homo neanderthalensis*, see Stringer, 1993.

231 For cultural overlap at Saint-Césaire, see Mercier et al., 1991.

238 Stringer and Gamble, 1993.

239 Ardrey, 1961.

The ten-foot spear was found in Germany amid the fossil ribs of an elephant; see Tattersall, 1993.

242 For the idea of speciation in space, see the final page of Stanley, 1981.

243 Galton, 1883.

For the Immigration Act of 1924, see Kevles, 1985, p. 97.

244 For the feats of Mozart and Mill, see Cox, 1926.

For Teller and Frisch, see Halberstam, 1994.

For details on sperm banks, see Kevles, 1985, pp. 259–64.

BIBLIOGRAPHY

Ackerman, Sandra, *Discovering the Brain* (National Academy of Sciences, 1992).

Ackersten, William A., "Canine Function in *Smilodon* (Mammalia; Felidae; Machairodontinae)," *Contributions to Science, Los Angeles Museum of Natural History* no. 356:1–22 (1985).

Altman, Stuart A., and Jeanne Altman, *Baboon Ecology: African Field Research* (University of Chicago Press, 1970).

Ardrey, Robert, *African Genesis: A Personal Investigation into the Animal Origins and Nature of Man* (Atheneum, 1961).

Arsuaga, Juan-Louis, et al., "Three New Human Skulls from the Sima de los Huesos Middle Pleistocene Site in Sierra de Atapuerca, Spain," *Nature* 362:534–37 (1993).

Berge, Christine, "How did the australopithecines walk? A biomechanical study of the hip and thigh of *Australopithecus afarensis*," *Journal of Human Evolution* 26:259–73 (1994).

Berger, L. R. and R. J. Clarke, "Eagle Involvement in Accumulation of the Taung Child Fauna," *Journal of Human Evolution* 29:275–99 (1995).

Bertram, Brian C. R., "Serengeti Predators and Their Social Systems," pp. 221–48 in *Serengeti: Dynamics of an Ecosystem*, A. R. E. Sinclair and M. Norton-Griffiths, eds. (University of Chicago Press, 1979).

Bishop, J. A., and L. M. Cook, "Industrial Melanism and the Urban Environment," *Advances in Ecological Research* 11:373–404 (1980).

Bonnefille, Raymonde, "Implications of Pollen Assemblage from the Koobi Fora Formation, East Rudolf, Kenya," *Nature* 264:403–7 (1976).

——, "Evidence for a Cooler and Drier Climate in the Ethiopian Highlands Towards 2.5 Myr Ago," *Nature* 303:487–91 (1983).

Bradshaw, John, and Lesley Rogers, *The Evolution of Lateral Asymmetries, Language, Tool Use, and Intellect* (Academic Press, 1993).

Brain, Charles K., *The Hunters or the Hunted? An Introduction to African Cave Taphonomy* (University of Chicago Press, 1981).

Broecker, W. S. "The Biggest Chill," *Natural History* 97:74–84 (1987).

Bunn, Henry T., and Ellen M. Knoll, "Systematic Butchery by Plio/Pleistocene Hominids at Olduvai Gorge, Tanzania," *Current Anthropology* 27:431–42, with comments from other scientists on pp. 442–52 (1986).

Carbonell, E., J. M. Bermúdez de Castro, J. L. Arsuaga, J. C. Díez, A. Rosas, G. Cuenca-Bescós, R. Sala, M. Mosquera, X. P. Rodriguez," Lower Pleistocene Hominids and Artifacts from Atapuerca-TD6 (Spain)," *Science* 269:826–30.

Cartmill, Matt, "Pads and Claws in Arboreal Locomotion," pp. 45–83 in *Primate Locomation*, Farish A. Jenkins, ed. (Academic Press, 1974).

——, *A View to a Death in the Morning: Hunting and Nature in History* (Harvard University Press, 1993).

Cavalli-Sforza, Luigi Luca, "Genes, Peoples, and Languages," *Scientific American*, 265:104–10 (1991).

Cerling, Thure E., "Development of Grasslands and Savannahs in East Africa During the Neogene," *Paleogeography, Paleoclimatology, Palaeoecology* 97:241–47 (1992).

Clark, J. D., et al., "African *Homo Erectus:* Old Radiometric Ages and Young Oldowan Assemblages in the Middle Awash Valley, Ethiopia," *Science* 264:1907–10 (1994).

Clarke, Ronald J. and Phillip V. Tobias, "Sterkfontein Member 2, Foot Bones of the Oldest South African Hominid," *Science* 269:521–24.

Cline, R. M., and J. D. Hays, "Investigation of Late Quaternary Paleoceanography and Paleoclimatology," *Geological Society of America Memoir* 145 (1976).

Clutton-Brock, T. H., "Size, Sexual Dimorphism, and Polygyny in Pri-

mates," pp. 51–60 in *Size and Scaling in Primate Biology*, William
L. Jungers, ed. (Plenum, 1985).

Coates, A. G., J. B. C. Jackson, L. S. Collins, T. M. Cronin, H. J. Dowsett,
L. M. Bybell, P. Jung, and J. A. Obando, "Closure of the Isthmus of
Panama: The Near-Shore Marine Record of Costa Rica and West-
ern Panama," Geological Society of America Bulletin 104:814–28.

Corballis, Michael C., *The Lopsided Ape: Evolution of the Generative
Mind* (Oxford University Press, 1991).

Count, Earl W., "Brain and Body Weight in Man: Their Antecedents in
Growth and Evolution," *Annals of the New York Academy of Sciences*
46:993–1122 (1947).

Cox, Catherine M., *The Early Mental Traits of Three Hundred Geniuses*
(Stanford University Press, 1926), pp. 594, 707.

Dart, Raymond A., "*Australopithecus africanus*: The Man-Ape of South
Africa," *Nature* 115:195–99 (1925).

———, "Innominate Fragments of *A. prometheus*," *American Journal of
Physical Anthropology* 7:301–38 (1949).

Dart, Raymond, A., with Dennis Craig, *Adventures with the Missing Link*
(Harper, 1959).

Darwin, Charles, *On the Origin of Species by Means of Natural Selection* (John Murray, 1859).

———, *The Descent of Man and Selection in Relation to Sex* (John Murray, 1871).

Dawson, Alastair G., *Ice Age Earth: Late Quaternary Geology and Climate* (Rutledge, 1992).

Deacon, Terrence W., "Rethinking Mammalian Brain Evolution," *American Zoologist* 30:629–705 (1990).

de Menocal, Peter B., William F. Ruddiman, and Edward M. Pokras,
"Influences of High- and Low-Latitude Processes on African Cli-
mate: Pleistocene Eolian Records from Equatorial Atlantic Ocean
Drilling Program Site 663," *Paleoceanography* 8:209–42 (1993).

Devore, Irving, and Sherwood L. Washburn, "Baboon Ecology and
Human Evolution," pp. 335–67 in *African Ecology and Human Evolution*, Howell F. Clark and Francois Bourliere, eds. (Aldine, 1963).

Diamond, Marian C., *Enriching Heredity: The Impact of the Environment
on the Anatomy of the Brain* (Collier, 1988).

Dobzhansky, Theodosius, *Mankind Evolving* (Yale University Press, 1962).

Donald, Merlin, *Origins of the Modern Mind* (Harvard University Press, 1991).

Durham, J. Wyatt and F. Stearns MacNeil, "Cenozoic Migrations of Marine Invertebrates Through the Bering Strait Region," pp. 326–29 in *The Bering Land Bridge*, David M. Hopkins, ed. (Stanford University Press, Stanford, California, 1967).

Eccles, John C., *Evolution of the Brain: Creation of the Self* (Routledge, 1989), pp. 215–16.

Edelman, Gerald, *Bright Air, Brilliant Fire: On the Matter of the Mind* (Basic Books, 1992).

Eldredge, Niles, "The Allopatric Model and Phylogeny in Paleozoic Invertebrates," *Evolution* 25:156–67 (1971).

Eldredge, Niles, and Stephen Jay Gould, "Punctuated Equilibria: An Alternative to Phyletic Gradualism," pp. 82–115 in *Models in Paleobiology*, Thomas J. M. Schopf, ed. (Freeman, Cooper, 1972).

Feibel, Craig S., et al., "Stratigraphic Context of Fossil Hominids from the Omo Group Deposits: Northern Turkana Basin, Kenya and Ethiopia," *American Journal of Physical Anthropology* 28:595–622 (1989).

Foley, Robert, *Another Unique Species: Patterns in Human Evolutionary Ecology* (Longman Scientific and Technical, 1987).

Frakes, Lawrence A., Jane E. Francis, and Jozef I. Syktus, *Climate Modes of the Phanerozoic: The History of the Earth's Climate over the Past 600 Million Years* (Cambridge University Press, 1992).

Gabunia, L., and A. Vekua, "A Plio-Pleistocene Hominid from Dmanisi, East Georgia, Caucasus," *Nature* 373:509–12 (1995).

Galton, Francis, *Inquiries into Human Faculty and Its Development* (Macmillan, 1883).

Gardner, R. Allen, and Beatrice T. Gardner, "Teaching Sign Language to a Chimpanzee," *Science* 165:664–72 (1969).

Ghiglieri, Michael P., *The Chimpanzees of Kibale Forest: A Field Study of Ecology and Social Structure* (Columbia University Press, 1984).

Goodall, Jane, *The Chimpanzees of Gombe: Patterns of Behavior* (Belknap Press, 1986).

Gould, Stephen Jay, *Ontogeny and Phylogeny* (Harvard University Press, 1977).

Gowlett, J. A. J., et al., "Early Archaeological Sites, Hominid Remains, and Traces of Fire from Chesowanja, Kenya," *Nature* 294:125–29 (1981).

Grine, Frederick E., ed., *Evolutionary History of the "Robust" Australopithecines* (Alden de Gruyter, 1988).

Halberstam, David, *The Fifties* (Ballantine, 1994), pp. 88–89.

Hamilton, A. C., *Environmental History of East Africa: A Study of the Quaternary* (Academic Press, 1982).

Hay, Richard L., *Geology of the Olduvai Gorge: A Study of Sedimentation in a Semiarid Basin* (Berkeley: University of California Press, 1976).

Holloway, Ralph L., "Human Paleontological Evidence Relevant to Language Behavior," *Human Neurobiology* 2:105–14 (1983).

Holmes, Arthur, *The Age of the Earth* (Harper and Brothers, 1913).

Holt, A. Barry, et al., "Brain Size and the Relation of the Primate to the Nonprimates," pp. 23–44 in *Fetal and Postnatal Cellular Growth: Hormones and Nutrition*, D. B. Cheek, ed. (John Wiley, 1983).

Hopkins, Brian, *Forest and Savannah: An Introduction to Tropical Terrestrial Ecology with Special Reference to West Africa*, 2nd ed. (Heinemann, 1974).

Howells, William W., *Getting There: The Story of Human Evolution* (Compass, 1993).

Hsü, Kenneth J., "When the Mediterranean Dried Up," *Scientific American* 227:27–36 (1972).

Huxley, Leonard, ed., *Life and Letters* (Appleton, 1900), pp. 192–204.

Hyman, Libbie H., *The Invertebrates: IV Echinodermata* (McGraw-Hill, 1955), pp. 462–67.

Imbrie, John, and Katherine Palmer Imbrie, *Ice Ages: Solving the Mystery* (Enslow, 1979).

Jerison, Harry J., *Evolution of the Brain and Intelligence* (Academic Press, 1973), pp. 8–9.

Johanson, D. C., et al., "New Partial Skeleton of *Homo habilis* from Olduvai Gorge, Tanzania," *Nature* 327:205–9 (1987).

Johanson, Donald, and Maitland Edey, *Lucy: The Beginnings of Humankind* (Simon and Schuster, 1981).

Johanson, Donald, and James Shreeve, *Lucy's Child: The Discovery of a Human Ancestor* (Morrow, 1989).

John, Brian S., ed., *The Winters of the World: Earth under the Ice Ages* (John Wiley, 1979).

Keith, Arthur, "The Taungs Skull," *Nature* 116:11 (1925).

Kennedy, Gail E., "A Morphometric and Taxonomic Assessment of a Hominine Femur from the Lower Member, Koobi Fora, Lake Turkana," *American Journal of Physical Anthropology* 61:429–36 (1983).

Kevles, Daniel J., *In the Name of Eugenics: Genetics and the Uses of Heredity* (Knopf, 1985).

Kingdon, Jonathon, *Self-Made Man: Human Evolution from Eden to Extinction?* (John Wiley, 1993), chapter 6.

Klein, Richard G., *The Human Career: Human Biological and Cultural Origins* (University of Chicago Press, 1989).

Kruuk, Hans, *The Spotted Hyena: A Study of Predation and Social Behavior* (University of Chicago Press, 1972).

Kurtén, Bjorn, *Pleistocene Mammals of Europe* (Aldine, 1968).

Kurtén, Bjorn, and Elaine Anderson, *Pleistocene Mammals of North America* (Columbia University Press, 1980).

Lanpo, Jia, and Huang Weiwen, *The Story of Peking Man* (Oxford University Press, 1990).

Leakey, L. S. B., P. V. Tobias, and J. R. Napier, "A New Species of the Genus *Homo* from Olduvai Gorge, Tanzania," *Nature* 202:308–12 (1964).

Leakey, M. D., and J. M. Harris, *Laetoli: A Pliocene Site in Northern Tanzania* (Clarendon Press, 1987).

Leakey, Meave G., Craig S. Feibel, Ian McDougall, and Alan Walker, "New Four-million-year-old Hominid Species from Kanapoi and Allia Bay, Kenya," *Nature* 376:565–71 (1995).

Leakey, Richard E., et al., "A Partial Skeleton of a Gracile Hominid from the Upper Burgi Member of the Koobi Fora Formation, East Lake Turkana, Kenya," pp. 167–73 in *Hominidae*, G. Giacobini, ed. (Jaca, 1989).

Lewin, Roger, *The Origin of Modern Humans* (Scientific American Library, 1993).

Lovejoy, C. Owen, "The Origin of Man," *Science* 211:341–50 (1981).

Lyell, Charles, *Principles of Geology*, vol. 3 (John Murray, 1833).

MacNeilage Peter F., et al., "Primate Handedness Reconsidered," *Behavioral and Brain Sciences* 10:247–303 (1987).

Marean, Curtis W., "Sabertooth Cats and Their Relevance for Early Hominid Diet and Evolution," *Journal of Human Evolution* 18:559–82 (1989).

Martin, Robert D., "Evolution of the Brain in Early Hominids," *Ossa* 14:49–62 (1989).

Marzke, M. W., "Joint Functions and Grips of the *Australopithecus afarensis* Hand, with Special Reference to the Region of the Capitate," *Journal of Human Evolution* 12:197–211 (1983).

Mayr, Ernst, "Taxonomic Categories in Fossil Hominids," *Cold Spring Harbor Symposia on Quantitative Biology* 15:109–18 (1950).

————, *Animal Species and Evolution* (Harvard University Press, 1963), p. 604.

McGrew, W. C., *Chimpanzee Material Culture: Implications for Human Evolution* (Cambridge University Press, 1992).

McKee, J. K., "Faunal Dating of the Taung Hominid Fossil Deposit," *Journal of Human Evolution* 25:363–76.

McHenry, Henry M., "The Capitate of *Australopithecus afarensis* and *A. africanus*," *Journal of Physical Anthropology* 62:187–98 (1983).

————, "Femoral Lengths and Stature in Plio-Pleistocene Hominids," *American Journal of Physical Anthropology* 85:149–58 (1991a).

————. "Sexual Dimorphism in *Australopithecus afarensis*," *Journal of Human Evolution* 15:177–91 (1991b).

Mercier N., et al., "Thermoluminescence Dating of the Late Neanderthal Remains from Saint-Césaire," *Nature* 351:737–39 (1991).

Montagu, Ashley, *Growing Young*, 2nd ed. (Bergin & Garvey, 1989).

Morin, Phillip A., et al., "Kin Selection, Social Structure, Gene Flow, and the Evolution of Chimpanzees," *Science* 265:1193–1201 (1994).

Nilsson, Tage, *The Pleistocene: Geology and Life in the Quaternary Age* (D. Reidel, 1983).

Ormaston, H., "Glaciers, Glaciation and Equilibrium Line Altitudes on Kilimanjaro," pp. 7–30 in *Quaternary Environmental Research on East African Mountains*, William C. Mahoney, ed. (A. A. Balkema, 1989).

Osborn, Henry Fairfield, *The Age of Mammals in Europe, Asia and North America* (Macmillan, 1910).

Owens, Mark, and Delia Owens, *Cry of the Kalahari* (Houghton Mifflin, 1984).

Posner, Michael I., and Marcus E. Raichle, *Images of Mind* (Scientific American Library, 1994).

Potts, Richard, *Early Hominid Activities at Olduvai* (Aldine de Gruyter, 1988).

Prothero, Donald R., *The Eocene-Oligocene Transition: Paradise Lost* (Columbia University Press, 1994).

Rightmire, G. Philip, *The Evolution of* Homo Erectus: *Comparative Anatomical Studies of an Extinct Human Species* (Cambridge University Press, 1990).

Rodman, Peter S., and Henry M. McHenry, "Bioenergetics and the Origin of Hominid Bipedalism," *American Journal of Physical Anthropology* 52:103–6 (1980).

Romer, Alfred S., *The Vertebrate Story* (University of Chicago Press, 1959), p. 327.

Ruff, Christopher B., "Climate and Body Shape in Hominid Evolution," *Journal of Human Evolution* 21:81–105 (1991).

Sarnthein, Michael, "Sand Deserts during Glacial Maximum and Climatic Optimum," *Nature* 272:43–46 (1978).

Savage, Thomas S., and Jeffries Wyman, "Notice of the External Characters, Habits and Osteology of Troglodytes Gorilla, A New Species of Orang from the Gaboon River," *Boston Journal of Natural History* 5:417–42 (1847).

Savage-Rumbaugh, E. S., *Ape Language: From Conditioned Response to a Symbol* (Columbia University Press, 1986).

Schaller, George B., *The Mountain Gorilla: Ecology and Behavior* (University of Chicago Press, 1963), pp. 169–98.

———, *The Serengeti Lion: A Study of Predator-Prey Relations* (University of Chicago Press, 1972).

Schrenk, Friedemann, et al., "Oldest *Homo* and Biogeography of the Malawi Rift," *Nature* 365:833–36 (1993).

Shipman, Pat, "Scavenging or Hunting in Early Hominids: Theoretical Framework and Tests," *American Anthropologist* 88:27–43 (1986).

Shipman, Pat, and Walker, Alan, "The Costs of Becoming a Predator," *Journal of Human Evolution* 18:373–91 (1989).

Sinclair, A. R. E., and M. Norton-Griffiths, eds., *Serengeti: Dynamics of an Ecosystem* (University of Chicago Press, 1979).

Skeat, Walter W., and Charles O. Blagden, *Pagan Races of the Malay Peninsula* (Macmillan, 1906).

Spencer, Frank, *Piltdown: A Scientific Forgery* (Oxford University Press, 1990).

Spoor, Fred, Bernard Wood, and Frans Zonneveld, "Implications of Early Hominid Labyrinthine Morphology for Evolution of Human Bipedal Locomotion," *Nature* 369:645–48 (1994).

Springer, Sally P., and George Deutsch, *Left Brain, Right Brain*, 5th ed. (W. H. Freeman, 1993).

Stanley, Steven M., "Paleoecology and Diagenesis of Key Largo Limestone, Florida," *Bulletin of the American Association of Petroleum Geologists* 50:1927–47 (1966).

———, "Functional Morphology and Evolution of Byssally Attached Bivalve Mollusks," *Journal of Paleontology* 46:165–212 (1972).

———, "A Theory of Evolution above the Species Level," *Proceedings of the National Academy of Sciences* 72:646–50 (1975).

———, *Macroevolution: Pattern and Process* (W. H. Freeman, 1979).

———, *The New Evolutionary Timetable: Fossils, Genes, and the Origin of Species* (Basic Books, 1981).

———, "Anatomy of a Regional Mass Extinction: Plio-Pleistocene Decimation of the Western Atlantic Bivalve Fauna," *Palaios* 1:17–36 (1986).

———, *Extinction* (Scientific American Press, 1987).

———, "An Ecological Theory for the Origin of *Homo*," *Paleobiology* 18:237–57 (1992).

———, "New Horizons for Paleontology, with Two Examples: The Rise and Fall of the Cretaceous Supertethys and the Cause of the Modern Ice Age," *Journal of Paleontology* 69:999–1007.

Stanley, Steven M., and William F. Ruddiman, "Neogene Ice Age in the North Atlantic Region: Climatic Changes, Biotic Effects, and Forcing Factors," pp. 118–33 in *Effects of Past Global Change on Life* (National Academy of Sciences, 1995).

Stern, Jack T., and Randall L. Susman, "The Locomotor Anatomy of *Australopithecus afarensis*," *American Journal of Physical Anthropology* 60:279–317 (1983).

Stringer, Chris, "Secrets of the Pit of the Bones," *Nature* 362:501–2 (1993).

Stringer, Christopher, and Clive Gamble, *In Search of the Neanderthals: Solving the Puzzle of Human Origins* (Thames and Hudson, 1993).

Susman, Randall L., Jack T. Stern, and William L. Jungers, "Arboreality and Bipedality in the Hadar Hominids," *Folia Primatologica* 43:113–56 (1984).

Sutcliff, Anthony J., *On the Track of Ice Age Mammals* (Harvard University Press, 1985), pp. 117–50.

Swisher C. C., et al., "Age of the Earliest Known Hominids in Java, Indonesia," *Science* 263:1118–21 (1994).

Tague, Robert G., "Variation in Pelvic Size between Males and Females," *American Journal of Physical Anthropology* 80:59–71 (1989).

Tattersall, Ian, *The Human Odyssey* (Prentice Hall, 1993), p. 127.

Theunissen, Bert, *Eugène Dubois and the Ape-Man from Java: A History of the First "Missing Link" and Its Discoverer* (Kluwer Academic Publishers, 1989).

Tobias, Phillip V., "The Brain of *Homo habilis*: A New Level of Organization in Cerebral Evolution," *Journal of Human Evolution* 16:741–61 (1987).

Trinkaus, Erik, and Pat Shipman, *The Neanderthals: Changing the Image of Mankind* (Alfred A. Knopf, 1993).

Turner, Alan, "The Evolution of the Guild of Larger Terrestrial Carnivores in the Plio-Pleistocene of Africa," *Geobios* 23:349–68 (1990).

Van Valkenburgh, Blaire, Mark F. Teaford, and Alan Walker, "Molar Microwear and Diet in Large Carnivores: Inferences concerning Diet in the Sabertooth Cat, *Smilodon fatalis*," *Journal of Zoology* (London) 222:319–40 (1990).

Vrba, E. S., G. H. Denton, and M. L. Prentice, "Climatic Influences on Early Hominid Behavior," *Ossa* 14:127–56 (1989).

Vrba, Elisabeth S., "Chronological and Ecological Implications of the Fossil Bovidae at the Sterkfontein Australopithecine Site," *Nature* 250:19–23 (1974).

———, "Some Evidence of Chronology and Paleoecology of Sterkfontein, Swartkrans, and Kromadrai from the Fossil Bovida," *Nature* 254:301–4 (1975).

———, "The Significance of Bovid Remains as Indicators of Environmental and Predation Patterns, pp. 247–71 in *Taphonomy and Paleoecology*, A. K. Behrensmeyer and A. P. Hill, eds. (University of Chicago Press, 1980).

Walker, Alan, and Richard Leakey, *The Nariokotome* Homo erectus *Skeleton* (Harvard University Press, 1993).

Walker, Alan, and Richard E. F. Leakey, "The Hominids of East Turkana," *Scientific American* 239:54–66 (1978).

Walter, R. C., et al., "Laser-fusion ^{40}Ar ^{39}Ar Dating of Bed I, Olduvai Gorge, Tanzania," *Nature* 354:145–49 (1991).

Walter, Robert C., "Age of Lucy and the First Family: Single-Crystal ^{40}Ar ^{39}Ar Dating of the Denen Dora and Lower Kada Hadar Members of the Hadar Formation, Ethiopia," *Geology* 22:6–10 (1994).

Webb, S. David, and Neil D. Opdyke, "Global Climatic Influence on Cenozoic Land Mammal Faunas," pp. 184–208 in *Effects of Past Global Change on Life* (National Academy of Sciences, 1995).

Wesselman, Henry B., "The Omo Micromammals: Systematics and Paleoecology of Early Man Sites from Ethiopia," *Contributions to Vertebrate Evolution* 7:1–219 (1984).

Whitmore, T. C., *Tropical Rainforests* (Clarendon, 1990).

Wood-Jones, Frederick, *Arboreal Man* (Hafner, 1964).

Woodward, Arthur Smith, letter under "The Fossil Anthropoid Ape from Taungs," *Nature* 115:235–36 (1925).

Yellen, John E., et al., "A Middle Stone Age Worked Bone Industry from Katenda, Upper Semliki Valley, Zaire," *Science* 268:553–56 (1995).

INDEX

Page numbers with illustrations appear in **_boldface italics._**

ABOUT THE AUTHOR

STEVEN M. STANLEY, Professor of Paleobiology at Johns Hopkins University, is a former Guggenheim Fellow. He has been elected to the National Academy of Sciences and the American Academy of Arts and Sciences and has served as president of the Paleontological Society. His previous books include *The New Evolutionary Timetable*, which was nominated for the American Book Award, *Extinction*, and four widely adopted textbooks. Although best known for his contributions to evolutionary theory, he also conducts hands-on research with fossils in the field and laboratory. He lives in Baltimore with his wife and ten-year-old daughter, recently adopted in Russia. He swims for exercise and gardens for relaxation.